About the Editor

MICHAEL K. DEAVER is the author of *Nancy* and the
bestselling *A Different Drummer: My Thirty Years with
Ronald Reagan.* One of Reagan's closest advisors, Deaver
served as assistant to the president and deputy chief of
staff in the White House from January 1981 until May
1985. He currently serves as vice chairman, international,
for Edelman Worldwide. He lives with his wife in Maryland.

ALSO BY MICHAEL K. DEAVER

Behind the Scenes

A Different Drummer: My Thirty Years with Ronald Reagan

Nancy: A Portrait of My Years with Nancy Reagan

WHY I AM A REAGAN CONSERVATIVE

EDITED BY

Michael K. Deaver

HARPER

NEW YORK • LONDON • TORONTO • SYDNEY

HARPER

A hardcover edition of this book was published in 2005 by William Morrow, an imprint of HarperCollins Publishers.

HarperCollins books may be purchased for educational, business, or sales promotional use. For information please write: Special Markets Department, HarperCollins Publishers, 10 East 53rd Street, New York, NY 10022.

FIRST HARPER PAPERBACK PUBLISHED 2006.

Designed by Kelly S. Too

The Library of Congress has catalogued the hardcover edition as follows:

Why I am a Reagan conservative / edited by Michael K. Deaver.— 1st ed.
    p. cm.
  ISBN 0-06-055976-4 (alk. paper)
    1. Conservatism—United States. I. Deaver, Michael K.

JC573.2.U6W538 2005
320.52'092'273—dc22

                                    2004061006

ISBN-10: 0-06-055977-2 (pbk.)
ISBN-13: 978-0-06-055977-9 (pbk.)

06 07 08 09 10 ❖/RRD 10 9 8 7 6 5 4 3 2 1

To Ronald Reagan, who changed
all of our lives

CONTENTS

CONTENTS

CONTENTS

CONTENTS

INTRODUCTION

In early 2004, the *New York Times* created a small stir by quietly assigning an experienced reporter to a new assignment: cover conservatives. Bill Keller, the *Times'* executive editor, appeared to be in a slight bind. After all, does the *Times* have a reporter working the "liberal" beat?

Regardless of intent, I think Big Media has finally owned up to the fact that conservatives are here to stay. The current political landscape is not some residue from the Reagan Revolution; it is the result of a generational sea change that will not dissipate anytime soon, despite the best efforts of Al Franken and Company.

When I first became active in the conservative movement of the Republican Party, we were considered a "fringe group," made up largely of westerners who were separated by far more than miles from the eastern GOP establishment.

I find myself bemused when today's liberals bemoan the rise of conservative medium outlets, identifying the usual suspects: Fox News, *Wall Street Journal* editorial writers, the *Washington Times,* Rush, and Sean Hannity.

They either forget—or chose to overlook—that liberal views dominated nearly every major institution for a good part of the century. From the think tanks to the National Council of Churches to the Democrat-dominated congressional committees to the newsrooms to the halls of academia, liberal views were the norm. National Public Radio's idea of a "balanced political dialogue" was—and still is—a pair of Democrats and a reporter from the *Washington Post.* Anybody to the right of Everett Dirksen was considered a menace or a lunatic.

Then a guy named Reagan came along.

The national week of mourning for Ronald Reagan reinforced the fact that conservatism, as we know it today, would not enjoy its majority status in American politics if not for him. Thirty-five million people tuned into the state funeral at the National Cathedral and the interment at the Reagan Library, more than had watched the final game of the NBA play-offs.

Reagan's impact on the American political scene is only rivaled by that of FDR in the twentieth century. Today the White House and both sides of Congress are in conservative hands. A majority of state legislatures are Republican. The majority of the Supreme Court is right of center.

I've never been viewed as a "movement conservative," but I've always believed in limited government, individual liberty, and the prospect of a strong America. These are principles that my friend Ronald Reagan espoused his entire life. The trick was his keen, unrivaled sense of timing and tone.

All of us who describe ourselves, for various reasons, as conservatives, stand on the shoulders of Ronald Reagan because he made conservatism respectable, acceptable, and now mainstream.

It is easy to survey today's political landscape and say the Gipper was right all along. But it could have just as easily failed. Reagan had his allies in this long national struggle. From its intellectual launch led by Bill Buckley and his merry band of "radicals," American conservatism today dominates because the Republican Party is the party of ideas. They may not all be the best ideas, but nobody looks to the Left for cutting-edge thinking.

Ideas radiate from think tanks like the Heritage Foundation, Hoover, Cato, and American Enterprise Institute, places that either didn't exist before 1980 or that the establishment overlooked. The tables have turned. Conservative intellectuals have muscled liberal think tanks like the venerable Brookings Institute to the side by pushing innovative tax cuts, school choice, and Social Security personal investment accounts.

Groups like Americans for Tax Reform and the National Federation of Independent Business are the interest groups to emulate. Nobody talks about being more like the National Education Association.

Liberals, long comforted by their dominance of the newsroom and editorial boards, have grown soft. Without new ideas, they can't take to the airwaves for three hours to defend a dogma that simply doesn't exist. The only people they are comfortable talking to are members of their dwindling constituency, a still-potent brew of special interests in the labor and environmental movements along with minority groups and academia.

What drives today's conservative? George W. Bush says without hesitation that he's a "proud conservative." Can John Kerry or future Democratic nominees for the presidency say, "I'm a proud liberal"? I doubt it.

THIS BOOK IS A COLLECTION OF ESSAYS BY SOME OF AMERICA'S most noted conservatives, the men and women who carry forward the mantle of Ronald Reagan. Each seeks to answer one simple question: Why am I a Reagan conservative?

—MICHAEL K. DEAVER
WASHINGTON, D.C.
OCTOBER 2004

WHY I AM A
REAGAN CONSERVATIVE

ROBERT L. BARTLEY

· ★ ·

BEING CONSERVATIVE GIVES YOU
A GRIP ON REALITY

Being conservative gives you a grip on reality. At least, that's what I think I've learned over my lifetime. Conservative principles and conservative approaches start in the real world as it exists, not in some lovely but imaginary utopia. While this certainly doesn't preclude reform and improvement, it does make you less likely to get carried away by the purely abstract.

I started political life as a Stassen Republican, I like to say, which back in those days meant a midwestern Republican with intellectual pretensions. That is, I wasn't much of a conservative. I missed the excitement of the Goldwater revolution, and I actually voted for Lyndon Johnson in 1964. In the heady one thousand days of the Kennedy administration, I was tempted by liberalism, thinking that finally we had a president who appreciated intellectuals, who was finally applying the university's insights to public life.

As I watched the 1960s unfold, I concluded that this was the height of folly. The intellectuals and the universities proved unwilling to enforce their bedrock principles, in particular freedom of speech. Supporters of the Vietnam War were not welcome or entitled to present their case, though that case was steeped in the containment policy, an intellectual creation by any standard. Drawing a line in Vietnam may very well have been a mistake; in retrospect, the way we fought the war clearly was. But it was a wave of self-righteous emotion, not a reasoned argument, that swept the campuses, the media, and liberalism generally. This was shortly repeated in toppling a president over Watergate; the tenor of the complaint was not that Richard Nixon broke some laws, but the fantastic notion that he was conducting a coup d'état.

I tried to apply the lessons of these experiences in more than thirty years of running the editorial pages of the nation's largest-circulation newspaper. Over this time I watched the conservative cause prosper. A few conservatives and neoconservatives were lonely in opposing détente, but we lived to see the Berlin Wall come down and the communist empire collapse. We were equally lonely in opposing wage-price controls and in supporting a cut in the capital gains tax in 1978, but we saw Ronald Reagan and Paul Volcker resolve the stagflation of the 1970s. When I started my career, Democrats were assumed to run the nation, but as the twenty-first century opened, George W. Bush was the president and Republicans controlled both houses of Congress. Mature thinkers always expect to be disappointed by their politicians, of course, but the swing toward conservatism is unmistakable.

Liberalism still clings to establishment power in the mainstream media and the academy. But conservatives are ascendant in the new media of cable television and the Internet. The universities wallow in deconstructionism and other flights of fancy, while conservatives built think tanks relevant to the real world. Conservatism has become not only the ascending political force, but the interesting intellectual one.

It was well put back in 1919 by that archconservative Rudyard Kipling. Let me end with a few stanzas of "The Gods of the Copybook Headings":

With the Hopes that our World is built on they were
    utterly out of touch.
They denied that the Moon was Stilton; they denied she
    was even Dutch.
They denied that Wishes were Horses; they denied that a
    Pig had Wings.
So we worshipped the Gods of the Market Who promised
    these beautiful things.

When the Cambrian measures were forming, They
    promised perpetual peace.

They swore, if we gave them our weapons, that the wars of
the tribes would cease.
But when we disarmed They sold us and delivered us
bound to our foe,
And the Gods of the Copybook Headings said: *"Stick to the
Devil you know."*

On the first Feminian Sandstones we were promised the
Fuller Life
(Which started by loving our neighbor and ended by
loving his wife)
Till our women had no more children and the men lost
reason and faith,
And the Gods of the Copybook Headings said: *"The Wages
of Sin is Death."*

In the Carboniferous Epoch we were promised abundance
for all,
By robbing selected Peter to pay for collective Paul;
But, though we had plenty of money, there was nothing
our money could buy,
And the Gods of the Copybook Headings said: *"If you don't
work you die."*

Then the Gods of the Market tumbled, and their smooth-
tongued wizards withdrew,

And the hearts of the meanest were humbled and began to
believe it was true.
That All is not Gold that Glitters, and Two and Two make
Four—
And the Gods of the Copybook Headings limped up to
explain it once more.

ROBERT L. BARTLEY, both a Pulitzer Prize and
Presidential Medal of Freedom recipient, was editor emeritus of the *Wall Street Journal* until his
death in December 2003.

ROBERT D. NOVAK

· ★ ·

GOVERNMENT: PROBLEM OR SOLUTION?

I grew up following my father's example as a liberal Republican. As a Korean War–vintage army officer who had been deeply influenced by Whittaker Chambers's *Witness,* I was a robust anti-Communist. Still, conservative writer John Chamberlain had it right in 1965 when he said I was a typical liberal Republican who thought government could be restrained and modified from its left-wing socialist model.

Chamberlain thought I was wrong, and he was correct. What really makes a conservative is whether you think the government always is the problem rather than the solution. I became a conservative in 1976 when I came to the conclusion—based on close observation as a reporter of how Washington worked—that it was the problem. Ronald Reagan and Calvin Coolidge are the only presidents of the twentieth and twenty-first centuries who agreed with that.

What makes me a conservative are the answers I give to

these questions: Can the government guarantee a stable economy? Is a high rate of taxation desirable to provide public "investment" in schools, infrastructure, and public services? Should the government serve as the arbiter choosing new technologies? Should the government protect industries by keeping out foreign competition?

To all these questions, I answer no. How many self-styled conservative politicians really reject these governmental nostrums?

ROBERT D. NOVAK writes the "Inside Report," one of the longest-running syndicated columns in the nation. He also appears on, and serves as co-executive producer of, CNN's political roundtable, *Capital Gang.*

BOB DOLE

· ★ ·

A LEGACY OF VALUES, NOT JUST A LABEL

To me "conservative" is a legacy of values that are at once timeless and vulnerable, not just a label. It is a faith; the humbling perspective that not every change represents progress; a fierce defense of individuals and national freedom; and a healthy skepticism toward institutions too large, too remote, and too impersonal to be truly democratic. Conservatives share the Founders' fears over too much power concentrated in too few hands. We prefer organizing society from the grassroots to dictating it from the top down.

Thomas Jefferson said, "The God who gave us life, gave us liberty at the same time." Growing up on the edge of the Depression-era Dust Bowl, I was taught to put my trust in God, not government, and never confuse the two. I eventually came to see conservatism as a creed of opportunity, rooted in the ability of seemingly ordinary people to accomplish extraordinary things. The worst of times brought out the best in my neighbors. In Russell, Kansas, adversity tested character. But it also bred a sense of responsibility for others who were hurting.

In any event, when I returned from World War II, I was sustained by neighbors who were anything but stingy with their love and encouragement. I learned then, if I hadn't already known it, that there is no such thing as a wholly self-made man or woman. Life has taught me well that the greatness of America lies, not in the power of the government, but in the goodness of her people. That's why genuine conservatives trust people to make their own decisions and realize their own dreams. We trust parents to choose the best education for their children. We trust entrepreneurs to generate new ideas and the jobs that follow. We entrust hard-earned dollars to the workers who earned them instead of centralized bureaucracies that limit options and frustrate dreams.

I confess that it took me years to fully understand conservatism and its many different interpretations. I was sometimes criticized as not being a "true" conservative by right-wing pundits and some one-issue special-interest groups. Being a compassionate conservative in the 1970s and 1980s was not appreciated by the right-wing ideologues, most of whom never ran for any office or cast a vote on any issue.

Of course, for a long time even genuine conservatives were the object of scorn, even ridicule from the Left . . . you know, we were lampooned as little old ladies in tennis shoes worried about Communists under the bed and fluoride in our water supplies, our overstuffed tycoons in batwing collars who were unwilling to look at the new moon out of respect for the old one. Ironically it was Ronald Reagan, the oldest of American presidents, who proved the most youthful of leaders. Far from living in the past, President Reagan looked forward to a future in which all of God's children were free, and all Americans celebrated the source of life and liberty.

Liberty, I might add, that should never be confused with license. Conservatives have no monopoly on virtue. Yet if we are true to our stated beliefs, we will take exceptions to a popular culture that all too often peddles trash for cash. Indeed, conservatives have a special responsibility, it seems to me—precisely because we embrace what President Reagan called the magic of the marketplace—to raise our voices in protest when the profit motive turns poisonous, coarsening our culture, polluting our air or airwaves.

In many ways my life traces the trajectory of American conservatism, from a marginalized faith in the bleak 1930s to triumph in the cold war to our current agenda-setting primacy. Recent tests have confirmed that the tide of events flows our way. In the days since 9/11 we have all drawn inspiration from young Americans, many in uniform far from home defending our most cherished values. You don't hear anyone questioning the courage or character of Generation

X these days. They know, just as surely as the heroes of Gettysburg or Omaha Beach, that freedom is never free. As a result, the world in 2004 is freer, more democratic, more entrepreneurial, and more hopeful than at any time in my life. That's why I remain what I proclaimed myself to be in 1996—the most optimistic man in America.

BOB DOLE, called "the most enduring Republican leader of the twentieth century," twice served as majority leader of the Senate. He currently serves as chairman of the International Commission on Missing Persons and of the National World War II Memorial.

MICHAEL BARONE

· ★ ·

DETROIT

Why am I a conservative? I usually answer that question with one word: Detroit. In the summer of 1967, after my first year at Yale Law School, I worked as an intern in the office of the mayor of Detroit. Politically, I was a liberal, an enthusiast for Lyndon Johnson's Great Society, and an admirer of the young liberal mayor, Jerome P. Cavanagh. Two years before, I had written an article in the *Harvard Crimson* contrasting him with Los Angeles's conservative mayor Sam Yorty: Yorty's policies had resulted in the Watts riot of 1965, I pointed out, while Detroit had had none. It turned out my analysis was premature. A riot broke out in Detroit one Saturday night in July 1967, and on Sunday I drove down from my parents' suburban home to the City-County Building to do what I could. The riot continued for an entire week, until federal troops were sent in. I was present in meetings in the police commissioner's office—what we called the Control Center—with the mayor and the governor of Michigan. When the sun went down around 9:00—this was Detroit's first summer with daylight savings time since World War II—the reports on the police radio would come crackling in. One square mile after another was being abandoned to the rioters, night after night. Whatever could be said for the liberal policies I favored, they had not prevented the deadliest urban riot of the 1960s.

I discovered other things when I worked at the mayor's office that summer. In June, the mayor told me to spend a week at the city's antipoverty program headquarters and interview the heads of all the divisions about what they were doing. The agency was headed by an energetic, optimistic man, but the division heads were an odd lot. Some were veterans of the social service bureaucracies, who seemed to be doing pretty much what they had been doing in their old agencies. One or two were clearly castoffs, people whom other agencies were glad to be rid of. Some were idealists pursuing innovative strategies. In school I had been enamored of the idea that city governments could transform the lives of their poor residents. Up close that seemed a much more dubious proposition.

The years after 1967 were horrible for Detroit. Crime soared, as Cavanagh's successor Coleman Young concentrated all his rhetorical fire on white suburbanites. Welfare dependency soared as well. The commercial frontage on the city's wide radial avenues became 99 percent vacant. Houses were abandoned, burned (especially on Halloween), gutted. In the neighborhood at the edge of the city where I lived between ages four and eleven, houses once empty were worth $3,000, less than the salvage value of their fixtures; residential real estate had become worthless. The population of Detroit is now half what it was when I entered kindergarten there in 1949. I know large parts of the city of Detroit block by block, and when I return there and drive around I am always appalled at what the liberal policies of the 1960s wrought or, at best, failed to prevent.

Detroit is an extreme case, but my liberal faith was also undermined by the nationwide economic malaise of the 1970s. When I took Economics 1 at Harvard in 1964, the lecturers and instructors made it clear that economic problems had been solved; they knew how to achieve permanent low-inflation economic growth. And markets, they insisted, had very little to do with it. The economy consisted of oligopolistic corporations, counterbalanced by large labor unions, under the supervision of government experts who could fine-tune supply and demand by manipulating a few numbers. Well, these confident people utterly failed to provide low-inflation economic growth in the 1970s. By then, a Democratic campaign consultant, I read the editorial pages of the *Wall Street Journal* and found I could not refute their arguments: low taxes and freer markets were a better way. Working on the podium of the Democratic National Convention, I watched the Kennedy campaign negotiating with the Carter campaign, allowing the Carterites to proceed on schedule in prime time in return for platform planks promising another $4 or $5 billion in subsidized public jobs. I decided I had to leave a profession I loved, and turned to journalism. Most journalists, I think, go into that line of work because they are liberals. I went into it because I was turning conservative.

I did not make the jump all at once, and I still don't press myself to take stands on issues on which I am torn. On some issues my positions have not changed: I favor free trade over protectionism, an activist foreign policy over isolationism, freer immigration rather than immigration restriction, a willingness to tolerate federal budget deficits rather than an insistence on fiscal tightness. And these are, after all, positions on which Ronald Reagan always agreed with Franklin Roosevelt; on those things, at least, Reagan was right when he said the Democrats had changed and he had not. As I see things now, we have moved in the last half-century from the industrial age to a postindustrial age, from centralization to decentralization, from an age when increasing bureaucratization seemed necessary to one in which a larger market sphere seems necessary. Industrial America naturally tended toward the liberal policies of half a century ago; postindustrial America naturally tends toward the conservative policies of today. Or so I have come to believe.

MICHAEL BARONE is a senior writer for *U.S. News & World Report* and has been the principal coauthor of *The Almanac of American Politics* since its first edition in 1972. He is also the author of *Our Country: The Shaping of America from Roosevelt to Reagan* and *The New Americans: How the Melting Pot Can Work Again.*

WILLIAM H. FRIST

· ★ ·

SELF-RELIANCE

The roots of a person's ideology can almost always be traced to their life experience—especially their early life experience. For me, there is no question that my conservative values come from the powerful role of family in my life.

My dad, in a letter he wrote to his great-grandchildren, stated plainly, "I am conservative. I believe the free enterprise system can do a better job at most things than the government can. People should learn to be self-reliant; when they are self-reliant, they will have self-respect."

Indeed, no idea is more central to the American dream than self-reliance—the idea that one rises on one's own merits. My life has been driven by the faith that every American can succeed with a strong sense of purpose, steady resolve, and steadfast commitment to the future.

Those values have had a direct impact on my life. My mother and dad, and my brothers and sisters, have all been driven to help others. After decades of education, training, and apprenticeship, I was able to achieve my dream of becoming a physician to improve people's lives through the healing arts.

I've been blessed with the opportunity to transplant human hearts—a procedure that was and remains on the frontier of medicine. Removing a heart from a life that had been lost and transplanting it to restore new life requires the latest surgical techniques, devices, and training.

It is this same new thinking and commitment to action that I also find appealing in conservatism. During my tenure in the Senate, conservatives have advanced and acted upon the ideas that have moved America forward while serving the cause of liberty we all hold so dearly in our hearts.

Conservatives have led the effort to reform welfare into a program that encourages dignity and independence. Conservatives have led the effort to reform education by holding schools accountable and raising student achievement. Conservatives have reduced the tax burden and helped restore prosperity during a difficult economic time.

All of these movements are to uplift the dignity of the individual so each can live a more fulfilling life.

Last, I am a conservative because I am a person of faith. I have always tried to live my life within an ethical and moral framework. My profession as a doctor reinforces my conviction that all life is precious and every child is born with the ability and the destiny to make a contribution to humanity.

Before I ran for the Senate, I didn't view the world through a certain ideology. I had no reason to. I was a full-time physician and viewed the world—as I still do mostly today—through a physician's eyes.

But when it came time to enter the political arena, there was no need to choose where I stood on the political spectrum. That's because—with my life experience—I already knew. So, perhaps I can offer an extended version of my dad's plain-stated truth: "I am conservative . . . because it is who I am."

SENATE MAJORITY LEADER WILLIAM H. FRIST, M.D., has served as a U.S. senator from Tennessee since 1994.

MARTIN ANDERSON

· ★ ·

A RINGING MELODY OF IDEAS

When I first became aware of Reagan in the 1960s, I was struck by how much I agreed with what he was saying. It all seemed so reasonable. No other political figure came close to laying out a policy blueprint that fit my personal convictions. When he talked about the Soviet threat, the danger of communism, the proper role of government, the need for low taxes, the greatness of our country, I just said to myself—"yes, yes, yes."

I don't think I ever became a Reagan conservative. As best I can remember I always was one.

But Reagan crystallized my political beliefs, as he did for millions of Americans and ultimately for millions of men and women all over the world. When he spoke, he spoke for me. He gave a powerful voice to what I wanted to see done. And the more I heard, the more I liked him. I never thought of him as an ideologue. It just seemed that what he advocated was reasonable, was sensible—was possible.

I suspect that was the secret to Reagan's political power. The political vision he carved out for himself fit the thoughts of many, many others. He did not persuade others so much as he rang a bell deep in their brains, a ringing melody of ideas that eventually swept across the United States and then spread to country after country.

He rarely argued his points; he just seemed to say, "Here is what I believe." In a sense he raised a standard, a vision of what he called "a shining city on a hill," a standard that appealed to hundreds, then thousands, and then millions. He seemed to be saying, "I am you, follow me." And they did.

Reagan governed the United States, the most powerful country on earth, for eight years. He left it far stronger than when he arrived. Today, some fifteen years after leaving office, the power of Reagan's ideas is even stronger than when he governed, and it continues to grow.

There is a special word for it—the noun *Reaganism.* That is an unusual mark in history. Few presidents (if any) leave such a powerful legacy. We say that some things are "Jeffersonian" or even "Lincolnesque." But we don't talk about Washingtonism or Lincolnism or Jeffersonism or even Rooseveltism.

We now have a clear idea of what Reagan accomplished. When he first walked into the Oval Office, the world was under the threat of an all-out nuclear war between the two superpowers—the Soviet Union and the United States. Communism was spreading throughout the world. Our economy was in tatters. It was an anxious, dangerous time.

When Reagan left office, eight years later in 1989, communism, not our economy, was in tatters. The cold war had been won. The threat of an all-out nuclear war that could have annihilated mankind had melted away. Our economy was surging ahead. And soon there would be only one superpower in the world—the United States.

How he accomplished all of this is still somewhat of a mystery to historians, but we are slowly beginning to unwrap the enigma of Reagan.

Perhaps most important, we have learned that the easygoing, smiling Reagan was not an "amiable dunce" as some believed. Behind the friendly facade was a cold brilliance, a razor-sharp mind that knew and understood far more than anyone suspected. When out of sight of his public, he was a compulsive worker, reading and writing and leaving a huge cache of handwritten documents that are now just being published for us to read. What made Reagan's accomplishments possible were many factors, but the two keys were a brilliant mind and tireless work.

Many other factors contributed to what Reagan accomplished. There was his love of liberty for all men and women, his sense of justice, an innate sense of what was right, his belief in God, his deep respect and liking for people of all races and religions, his enthusiasm for fun and adventure, his extraordinary health, and his intelligent, loving wife, Nancy, who was his rock of support.

But there was one other factor that few men have. You can be smart, have good ideas, be friendly and likable, and have all the other attributes that Reagan had, but if you don't have the courage and toughness to stick with what you believe, no matter what happens, you will never have a chance of doing what Reagan did.

Reaganism has become a political model that is used by many politicians in America, and even in other countries. When faced with a difficult problem, many politicians quietly say to themselves: "What would Reagan do?" I think we would be surprised at how often the answer is obvious.

———

MARTIN ANDERSON is the Keith and Jan Hurlbut Senior Fellow at the Hoover Institution. He was a senior policy adviser to the 1976 and 1980 presidential campaigns of Ronald Reagan and served as chief domestic and economic policy adviser under President Reagan.

DAVID A. KEENE

· ★ ·

THE FOUNDERS HAD IT RIGHT

Like many conservatives of my generation, I began life as a Democrat. My father was a labor union organizer and my mother once served as the president of the Women's Auxiliary of the United Auto Workers. As the faithful young son of good Democrats, my earliest political activity involved standing in the snow passing out literature for John F. Kennedy in the days leading up to the crucial 1960 Wisconsin Democratic Primary.

But, frankly, it didn't take. I was an avid reader and quickly discovered *National Review* and Friedrich Hayek. In fact, the most influential book I read in high school was not Hayek's *Road to Serfdom*, but his *Constitution of Liberty,* which was ordered by mistake and given to me by the librarian because she was told not to put it on our high school library's shelves. I still have it.

I soon came upon *Conscience of a Conservative* and became a rabid Goldwater fan. There weren't many of us at the University of Wisconsin, but we were loud and very active. We published the first university-based student conservative journal, and I became its editor.

Few of us got involved in the movement for opportunistic reasons. Those with an eye to "the main chance" were looking elsewhere. Bill Buckley, after all, likened our mission to standing on the tracks before an oncoming train and demanding that it stop. And Whittaker Chambers in *Witness* suggested pessimistically that in abandoning collectivist Communism he was abandoning the "winning" for the "losing" side.

The fact is, conservatives were reviled in those days as cranks or dangerous extremists. The party that eventually nominated Barry Goldwater and elected Ronald Reagan was firmly in the hands of an establishment that wished only to manage the continued growth of an already bloated government and wanted nothing more than to coexist with what Reagan later described so accurately as an "evil empire."

We joined up because we firmly believed that the Founders had it right and that both major parties were toying with policies that could ultimately cost us our freedoms at home and force a surrender to tyranny abroad. We knew we were right and were attracted to a movement that encouraged us to say what we believed and to fight for politicians who shared our values.

The conservative movement in those days was in its infancy. My mentors were Walter Judd and *National Review*'s Frank Meyer. Meyer would summon me to his mountain home in New York to make sure I was devoting myself to the "cause" and Judd would warn me not to neglect my studies while doing so.

I dropped out of school in 1964 to campaign for Barry Goldwater, became active in Young Americans for Freedom, ran unsuccessfully for the state legislature in 1969 in a campaign that featured radio ads cut for me by then California governor Ronald Reagan, moved to Washington the next year to work for Vice President Spiro Agnew and never looked back.

By 1972 my parents were voting Republican and not just so that I could keep my job. The country was changing. We began as what would later be known as Reagan Democrats, rejected the foreign and defense policies of the Democratic Party, and came to understand that liberty and security are often in conflict. Millions of former Democrats and liberals with similar backgrounds were following more or less the same path in the 1970s and, together, we, the ideas we championed, and the candidates for whom we toiled have managed to have a profound impact on the politics and future of the nation.

We haven't won, of course, but we've made a great deal of progress. History hasn't and isn't about to end. Frank Meyer believed, like Jefferson, that man's struggle for freedom had to be fought every day. My own politics have been based on a belief that the world has always been divided into two basic camps. The first is made up of those who believe that they know best how everyone should conduct their lives, order their business, and educate their children and see government as a way to impose their views on the less enlightened. The other camp always and everywhere is made up of those who just want to be left alone to arrange their own lives, earn an honest living, and raise their children as they believe they ought to be raised. They are skeptical of those who tell them they ought to live differently "for their own good" and leery of a government under the control of such people.

This country was founded to guarantee the rights and freedoms of those in that second camp and as conservatives we can be proud of our role in protecting those rights and freedoms.

DAVID A. KEENE is chairman of the American Conservative Union, the country's oldest and largest grassroots conservative organization. He is also a lobbyist with The Carmen Group, a governmental affairs and legislative relations firm based in Washington, D.C.

MICHAEL MEDVED

· ★ ·

CONSERVATISM: IT WORKS FOR ME

Your request for an explanation as to "Why I Am a Reagan Conservative" arrived, conveniently enough, in the same week that an ambitious new study on the origins of conservatism appeared in *Psychological Bulletin,* the official journal of the American Psychological Association.

The research project bore the intimidating title "Political Conservatism as Motivated Social Cognition" and used data from eighty-eight different surveys to argue that conservatives are troubled, authoritarian personalities who favor simplistic approaches to all life's challenges. The executive summary of this scholarly tour de force specifies that conservatives suffer from "death anxiety," "dogmatism," "intolerance of ambiguity or uncertainty," a strong "need for order, structure, and closure," no "openness to experience," "low self-esteem," and limited "integrative complexity." This analysis, proffered by distinguished psychology faculty at Berkeley, Stanford, and the University of Maryland, in effect defined conservative ideology as a form of mental illness. If you admire Ronald Reagan and George W. Bush, in other words, you ought to consider yourself one sick puppy—and a simpleminded one, at that.

When I discussed this psychological analysis on my syndicated radio show, I made the point that it actually revealed far more about the liberal mind-set than it does about the thinking or emotional health of American conservatives. Establishment liberals feel so certain that their ideology represents the only valid response to reality that they diagnose any dissent from their ideas as a form of psychological dysfunction. All those who question the leftist pieties they cherish must be motivated by greed, selfishness, cruelty, blindness, authoritarianism, or sheer stupidity.

Of course, this argument utterly ignores the most common justification that real-life conservatives provide for their political and cultural approach: like most of my colleagues, I embrace conservative ideas because they work. The values emphasized by right-wing thinkers and leaders—values of individual responsibility, fair competition, optimism, rewards for hard work, family, fidelity, religious faith, peace through strength and patriotism—maximize any individual's chances of achieving personal success and satisfaction. On the other hand, the tendencies of contemporary liberalism—whining, politically correct conformity, competitive victimization, dependence on government, sexual and social experimentation, emphasis on group identity over individualism, justification of criminality, militant secularism, and utopian pacifism—increase the likelihood of disaster and despair on both a national and individual basis.

My wife, an author and distinguished clinical psychologist (who resigned some years ago from the American Psychological Association, thank God), makes the point that most successful people in the United States choose to live conservative values in their private lives, regardless of their political orientation. Even University of California professors who define their conservative opponents as fearful and warped will rarely commit themselves on a personal basis to the radical notions they espouse in their work. From their elegant mansions in the Berkeley hills, these prominent academics may rant about the need for redistributing wealth or breaking down the tyranny of patriarchal marriage. But when it comes to redistributing their own luxury cars and fancy computers, or assigning their daughters (and potential grandchildren) to communal living arrangements that dispense with conventional marital and middle-class values, these advocates of brave new worlds will seldom live up to the logic of their public pronouncements.

Like most Americans who have achieved any measure of success, liberal opinion leaders reached their positions of influence through toil and competition and self-discipline, not through self-pity, complaint, indulgence, or placing ethnic identification above individual achievement.

Unlike the majority of liberal mandarins, I remember—and honor—the virtues and strategies that enabled me (with God's help) to earn the blessings that enrich my life. I don't feel guilty or apologetic for my pleasant home, my trophy wife (of nearly twenty years), our terrific kids, or my exciting and rewarding job. I believe that the same formula that has rewarded me can work for the overwhelming majority of Americans. A guilty leftist may fret that only those with unique talents, or privileged by lucky accidents of birth, can enjoy fortunate lives. Conservatives understand, however, that success stems from the unprecedented blessings showered by Providence on this nation as a whole, and from the middle-class virtues that achieving people seem to exemplify, often in spite of themselves.

In other words, it ought to be obvious that traditional values and conservative, free-market ideals have worked well for influential individuals—right, left, and center. I am a conservative because I believe that those time-honored habits of thought and behavior not only function to improve our private situations, but also serve to inevitably and simultaneously uplift the life of the nation.

MICHAEL MEDVED is a film critic who hosts a nationally syndicated daily radio talk show on the intersection of politics and pop culture. He is the author of *Hollywood vs. America The Shadow Presidents,* and seven other nonfiction books.

RICK SANTORUM

· ★ ·

BUILDING ON WHAT IS GOOD FOR AMERICA

I grew up in a traditional Roman Catholic family in Pittsburgh in the late 1960s and early 1970s, a product of the Vietnam era. Both my parents worked for the Veterans Administration, and they introduced me to many veterans as a young person. Knowing these veterans and the sacrifices they made for our country, I valued patriotism and respect for those who served.

I found conservative Republicans aligned with these values and liberal Democrats aligned with the 1960s protesters. Conservatives wanted to build on what is good for America. Americans are a great people with a great culture. During the 1960s revolution, liberals tried to tear down that culture. Like many other young conservatives, I became a disciple of Ronald Reagan, the politician who embodied the positive vision of America's future.

Faith and family played a big part in my becoming a conservative and running for elected office. Coming from a traditional Roman Catholic family, my parents engrained in me a love of traditional values. They taught me that people are called to serve, not to wage war on the government. I responded to that call to service when I ran for Congress fifteen years ago and continue to serve in the United States Senate today.

My philosophy remains decidedly conservative. I believe that the greatness of America lies in its citizens and in its families, and that what is best for our country is limited government and a strong defense.

RICK SANTORUM is a member of the Senate leadership, chairing the Senate Republican Conference. He has served the people of Pennsylvania in the Senate since 1995.

TRENT LOTT

· ★ ·

THE GOVERNMENT CLOSEST TO THE PEOPLE

I am a conservative because I feel as Thomas Jefferson did, that the best government is the government closest to the people and that the federal government should be held in check wherever possible.

I believe in fiscal responsibility, which means, "Don't spend it, if you don't have it."

I believe in a strong national defense because all our other freedoms and liberties could be lost without that.

And, finally, I have faith in individuals to assume responsibility for their own lives and not look to the government to solve all their problems.

TRENT LOTT has served four decades as a senator from Mississippi. Senator Lott also served as the Republican majority leader during the first two years of President George W. Bush's administration.

EDWIN MEESE III

· ★ ·

A JUST AND PROSPEROUS SOCIETY

I am a conservative because conservatism is the political philosophy that best addresses the moral values of freedom and responsibility and because it has proved to be the best foundation for a just and prosperous society. Conservatism understands the need for balance between liberty and order, and it has respect for tradition and experience as the basis for a successful civilization and for genuine human progress.

The characteristics of a conservative society include individual liberty, limited government, and free-market economics. Individual liberty involves freedom of religion, freedom of speech and advocacy, freedom of association, and freedom to participate in the political process.

Limited government, in which those in authority act by consent of the governed and are bound by the rule of law, is necessary to preserve those liberties and to avoid arbitrary, capricious, and unfair acts of officialdom. Limited government also reserves to the individual and to private voluntary associations the maximum control over their own decisions consistent with the needs of society as a whole.

Free-market principles facilitate efficient economic decision making that maximizes the potential for success, as opposed to the flawed "command and control" systems of a socialist economic regime.

These conditions provide the greatest opportunity for people to achieve their true potential and to take part in the political decisions that govern the society in which they live.

Conservatives know the value of history as a means of avoiding the mistakes of the past and as a means of understanding the human condition. Russell Kirk has written that "conservatives respect the wisdom of their ancestors" and "that the essence of social conservatism is preservation of the ancient moral traditions of humanity."

Conservatism best preserves the values that guided the Founders of the United States and protects the principles and system of government that they enshrined in the Constitution. The concept of a government limited to the powers enumerated in a written constitution, with such power divided among three separate branches and with checks and balances among the branches, was deemed necessary to prevent oppression and to protect freedom. Today it is the role of conservatives to initiate and support policies and political action in furtherance of those precepts.

The conservative philosophy promotes progress in society, but in a balanced and prudent way. Again to quote Kirk, "Conservatism is not a fixed immutable body of dogmata; conservatives inherit . . . a talent for re-expressing their convictions to fit the time," but are "opposed to the narrowing uniformity, egalitarianism, and utilitarian aims of most radical systems. . . . Custom, convention, and old prescription are checks both upon man's anarchic impulse and upon the innovator's lust for power." These are the conditions under which "society must alter, for prudent change is the means of social preservation."

During my lifetime I have seen the success of conservative philosophy prove its value to the nation and the world. The era of 1930 to 1950 saw the rise of socialism around the globe and, in the United States, the vast expansion of power in the federal government, fueled by reaction to the Great Depression of the 1930s and by the massive military buildup of World War II.

It was a concern about the drift toward socialism that inspired great thinkers like Friedrich von Hayek and Milton Friedman to nurture conservatism as a fledgling "intellectual" movement. In the 1950s, the advent of *National Review* and the discussions and writings of academics and a few journalists helped to popularize conservative ideas, particularly in the area of economic freedom.

In 1964, the presidential campaign of Barry Goldwater transformed conservatism into a "political" movement.

Though Goldwater himself was unsuccessful in the election, conservative ideas stimulated thousands of young people to become active in politics and to embrace concepts of freedom and limited government.

Ronald Reagan, first as governor of California and then as president of the United States, turned conservatism into a "governing" movement. His leadership and advocacy demonstrated that conservative principles worked in practice and promoted economic growth, individual freedom, and political success. This era changed the parameters of policy debate and left conservatism as the dominant political philosophy of the nation.

This does not mean that Leviathan has been conquered. Government is still too large and too consuming of resources at all levels. Diversity of thought is still negligible on too many higher education campuses, as leftist philosophy dominates academic circles. Much of the political establishment is still gripped by the "lust for power" that Russell Kirk decried.

But conservatism has proved its worth, has a firm intellectual foundation, and provides the basis for hope and the formula for action to achieve a better civilization in which freedom, opportunity, prosperity, and civil society flourish.

EDWIN MEESE III, former U.S. Attorney General under Ronald Reagan, is the Ronald Reagan Distinguished Fellow in Public Policy at The Heritage Foundation. Mr. Meese is also a Distinguished Visiting Fellow at the Hoover Institution, Stanford University, and a Distinguished Senior Fellow at the Institute of United States Studies, University of London.

J. C. WATTS, JR.

· ★ ·

REMEMBERING RONALD REAGAN

I remember a great sketch on *Saturday Night Live* a few years ago—back in its heyday—where Phil Hartman portrayed Ronald Reagan in the White House in a way most of us hadn't envisioned him. Actually, there were two Reagans in this comedy bit: the kindly, grandfatherly Reagan whom we all knew, greeting Jimmy Stewart and a Girl Scout in the Oval Office, followed by an autocratic Reagan, exiting into a side room to micromanage serious affairs of state. It was truly funny, because none of us ever imagined Reagan in the way he was portrayed in that situation room.

The brief portrayal of Jimmy Stewart in this sketch reminded me of *It's a Wonderful Life*. It made me pause to think of what my life and America would be like if Ronald Reagan had never lived . . . and that's a thought I'd rather not dwell on for long.

I was not a Republican when Ronald Reagan was president, but he sure got my attention. I guess you could call me a Reagan Democrat at the time. I was in my twenties, and like many Democrats, I did a gut check in the 1980s when President Reagan put to words the convictions of my heart. In many ways, President Reagan inspired me to become a Republican.

At first, Reagan appealed to that athletic, competitive side of me, more than the political side of me. I always thought that he could have rivaled Barry Switzer at halftime of an OU-Texas game. "We're down by a few points, but we're going to come back," he might have said in such a setting. You'd feel that it would happen, because he said it. You had a faith that good things were going to happen. His confidence was that infectious. He appealed to my hopes, not my fears. My confidence, not my doubts.

It's tough to execute a vision if you cannot communicate it. Candidate Reagan had a clear, most unambiguous vision of where he wanted to take our nation, and he communicated it with equal, unprecedented clarity. President Reagan executed that vision by bringing Americans with him on his journey. He delivered peace through strength, respect for the sanctity of life, and economic vitality by unleashing that entrepreneurial spirit of Main Street America. He got government off our backs.

But beyond that, President Reagan cultivated my homegrown values through his message of hope and economic opportunity for all. He rallied America around our highest common denominator, not our lowest. At a most basic level, he challenged America to think about what it wanted to be when it grew up. He envisioned a world free of communism and a nation alive with entrepreneurship. He was the kind of leader who looked toward the next generation, not the next election.

Ronald Reagan possessed a common sense that spoke to the common man. He had that kind of Buddy Watts–style reasoning that my dad shared with me as a kid growing up in Eufaula, Oklahoma. He believed in America's people. I always felt he'd be as comfortable and at home walking the streets of beautiful downtown Eufaula, Oklahoma, as he was in L.A. or D.C. Probably more so.

My dad, Buddy Watts, finished the sixth grade and spent two days in seventh grade. He never dined with ambassadors, kings, or presidents as I have been blessed to. He didn't quote Thoreau or Shakespeare. But there was no one smarter than Buddy Watts. Reagan possessed that same kind of wisdom, and it attracted millions of Americans like me. I wasn't attracted to Ronald Reagan because of his intellectual solutions . . . but because of his practical solutions.

When we lost our friend early in the summer of 2004, I didn't mourn President Reagan's death. Rather, I celebrated his life. While our nation continues its journey toward that "shining city on a hill," there is no doubt in my mind that President Reagan is now there watching, head cocked to the side with that signature infectious smile, pulling for that great goodness that is the place others call America, and we call home.

President Reagan often closed his speeches with the words "God bless you, and God bless America." God answered that prayer—for America and the world—by giving us the gift of Ronald Reagan.

All honest Americans—even those who may have had policy differences with him—will acknowledge our late great President's uplifting demeanor and optimistic prose. Perhaps my favorite Reagan quote is this exhortation: "Let us renew our determination, our courage, and our strength. And let us renew our faith and our hope. We have every right to dream heroic dreams."

Ronald Reagan was a heroic figure who empowered us not to just dream heroic dreams, but to see them through to reality. I thank God for Ronald Reagan.

The people in fictional Bedford Falls learned what life would have been like without George Bailey. I'm thankful for Ronald Reagan and can't imagine the world today without him.

J. C. WATTS, JR. was a U.S. representative from Oklahoma.

BILL OWENS

· ★ ·

OPTIMISM

Why am I a conservative? One word: optimism.

Conservatism is often caricatured—unfairly—as so at home in the past that it fears the challenge of an uncertain future. It is, however, our harnessing of enduring, first principles of freedom, opportunity, strength, and hard work that give us the confidence to embrace the future with certainty and enthusiasm.

It was Winston Churchill who said, "The farther back you look, the farther forward you see." As Americans, we look back to the principles of our nation's founding, including the bedrock idea that our inalienable rights include "life, liberty, and the pursuit of happiness." It is a system that thrives through free minds and free markets and in the dignity and the potential of every human being.

America has succeeded, and continues to thrive, not because of the depth and breadth of our government, but because of the drive and the determination of Americans to build better lives for themselves and their children. That fundamentally conservative vision is the heart of the American Dream, and it is filled with optimism.

In our nation, we believe that each person can reach higher and grasp success. Alexis de Tocqueville underscored this when he wrote, "While democracy seeks equality through liberty, socialism seeks equality through restraint and servitude."

Conservatism also works to preserve these democratic ideals through strength, understanding, as George Washington noted, that "to be prepared for war is one of the most effectual means of preserving peace." A confident America at home in the world is one of the best guarantors of freedom at home and around the globe.

America is a land of possibility because it is rooted in optimistic principles that have endured for more than two centuries. We conservatives proudly embrace these values as we look forward to a new century of promise and achievement. That is why America is, and always will be, President Reagan's "shining city on a hill."

BILL OWENS is governor of Colorado.

JON KYL

· ★ ·

SORTING OUT CONFLICT

Nearly everyone, whether a conservative, a liberal, or neither, contemplates at some point or other what it means to be human. My sense of what human nature is began with what I could observe: that conflicts were inherent in the human condition. I came to believe—and still believe—that conservatism is the philosophy that gives us the best chance of sorting out these conflicts.

I hasten to add that I haven't thought these things through entirely on my own over the last sixty-two years. Understanding came to me, in part, through listening to my father, to my mother, and to other members of our family who had very practical, commonsense answers to our problems.

The frugality and the dignity of my immigrant grandparents made a big impression on me. My family, my religious faith, my early experiences—all of these brought home to me that two things were very important: responsibility and self-reliance. And these also happen to be two qualities that I consider to be at the heart of American conservatism.

Education is a part of this, too, of course. I had a good education. I read what the American Founders wrote about these matters, and, again, it made an instinctive kind of sense to me. Jefferson and the Federalist Papers authors probed the nature of man. They pointed out that all of us are created by God, and that this is where our rights come from. Once you embrace this view of things, you see that government—which has power over individuals' rights and property—must be constrained if it is to be just. Our governmental structures are there to settle the conflicts that, as I said, arise between human beings unavoidably and in every generation. But government itself is not exempt. Government itself can go wrong, which argues for it to be limited. Arrogant, paternalistic government or the opposite problem, the breakdown of governance if individuals willfully pull away from the body politic in pursuit of their particular interests—these were the two dangers that the Founders were attempting to avoid, for the first time in the rather discouraging history of governments.

"Nearly all men can stand adversity, but if you want to test a man's character, give him power." So said Lincoln.

Power does corrupt—I've seen it. With these words, Lincoln indicates something about our government as it was set up under the Constitution. Because it is a limited government obligated to respect the God-given rights of citizens, those who are elected to represent the citizens have to resist the human temptation to take too much power unto themselves. Power rests ultimately with the people. So again, I'd say that Honest Abe, in clarifying the moral and political duties of officeholders in a democracy based on equality of rights, is a good conservative as I construe that term.

The twentieth-century book that first captured these ideas for me was Barry Goldwater's *The Conscience of a Conservative* (1960), which I devoured from cover to cover as a college freshman. Barry, who was later to become a mentor, articulated in his book what politics is, or should be, all about. He called it "the art of achieving the maximum amount of freedom of individuals that is consistent with the maintenance of social order." There is a lot of balance in that statement, despite what the opposing presidential campaign (Lyndon Johnson's) said about Goldwater in the famous 1964 race. Goldwater and the others I've spoken of in this essay courageously looked at the world as it is and tried to find a way to reconcile the many interests, the seething conflicts, the differing needs, and the diverging desires of citizens in our vast and complex democracy.

Conservatism is by no means the political cure-all for the

problems of human nature. Those will always be with us. And that's the point: while liberals may attempt to change human nature, conservatives try to constructively accommodate it. And that's what makes conservatism practical and realistic as well as ideal.

———————

JON KYL is a U.S. senator from Arizona.

HENRY J. HYDE

· ★ ·

THE IMPORTANCE OF TRANSCENDENT
MORAL VALUES

I'm a conservative because conservatives understand the importance of transcendent moral values in American public life.

Conservatives understand that democracy isn't a machine that can run by itself. Conservatives understand that it takes a certain kind of people, living certain virtues, to make self-governance work. Conservatives understand that politics is accountable to universal moral truths that stand in judgment on how we make our democracy work.

Conservatives are unashamed to speak about God, and the imprint of God on our public life. Conservatives understand that God inscribed a moral code on the human heart long before that code was chiseled onto tablets of stone. And conservatives understand that anyone who wants to live freedom nobly must attend to that moral code.

When Abraham Lincoln asked at Gettysburg whether a nation "so conceived and so dedicated" could "long endure," he was speaking to every generation of Americans, not just the generation of the Civil War. He was asking every generation of Americans whether they believed in, and were prepared to defend, the "self-evident" moral truths on which American democracy was founded. Yes, we have to take care that the structures of democratic government in America remain supple and responsive. What conservatives understand is that we must also, and always, take care to "conserve" the values, the moral truths, which are the foundation stones of American democracy.

It's because of these convictions that I have allied myself with the pro-life movement for thirty years. The pro-life movement in the United States is determined to "conserve" one of the basic moral truths on which American democracy rests: that every human being, from conception until natural death, has an inalienable right to life. In defending the unborn, however, pro-life Americans are "conserving" and defending a noble idea of America as a community of inclusion, in which every human being, regardless of condition, is welcomed in life and protected in law. That's the American way. That's the story of America—a story of broadening, not constricting, the community of common protection and concern. Conservatives understand that. And that's another reason why I'm a conservative.

I'm also a conservative because I'm a great believer in the people and their wisdom. To vary from my friend William F. Buckley, I would much rather be ruled by the first two hundred names in the DuPage County phone book than by the faculty of the University of Chicago, intelligent as they may be. I believe there is an inherent sense of justice in the people, a sense of justice that gives all of us the capacity for self-governance. I believe that the people's natural communities—their families and neighborhoods and clubs and small businesses and parishes—are the real schools of freedom in America. And I believe with Edmund Burke that those "small platoons" are crucial building blocks of democracy.

Finally, I'm a conservative because I believe America is worth defending. We are not perfect; no conservative would ever claim perfection for a political community. But with all our flaws, the United States is the embodiment of the noblest political aspirations of mankind. Those who wish to destroy the great American democratic experiment must be resisted: because they seek to destroy our homes and our neighbors, yes, and also because they seek to destroy an idea—the idea of free men and women governing themselves in justice and living civility amid plurality. That is what conservatives are defending in the new world war in which we're engaged.

A congressman since 1975, HENRY J. HYDE serves Illinois's Sixth District and is chairman of the House International Relations Committee, where he plays a key role in the war on terrorism. He also serves on the House Judiciary Committee.

G. GORDON ROBERT LIDDY

· ★ ·

KEEPING YOUR DREAM

I am a conservative for many reasons, but they can best be summarized by something told to me by the late, great actor James Cagney. I was a guest in his home years ago and he told me that when he was a boy, they were so poor, he and his brother slept in the same bed together and would both "scheme" at night about how they were going to get something to eat for breakfast. They also talked about their dreams. James Cagney's dream was to become a "hoofer," meaning a splendid dancer, which he did become in addition to being a fine actor. His brother's dream was to become a physician. This was, he said, "in the days before government programs, when people had only themselves upon which to rely."

Mr. Cagney said he achieved his dream and so did his brother, who became a prominent doctor. They did it on their own. Tellingly he said, "You know, Gordon, today with the welfare system, with one hand they give you the check and with the other hand they take away your dream."

G. GORDON ROBERT LIDDY is host of *The G. Gordon Liddy Show,* a syndicated, conservative radio talk show. He has written two novels and an autobiography and is an actor in motion pictures and television.

JIM DeMINT

THE STRUGGLE FOR FREEDOM

I often visit Hilton Head in my home state of South Carolina. One of my favorite things to do when I'm there is to get up early, go down to the South Beach Marina, and walk along the boardwalk by the quaint stores looking at the boats and just hanging around for a few minutes enjoying the morning with a cup of coffee.

One particular morning, I spotted a windsurfing class of about a dozen kids on the opposite side of the bay on a small beach along the inlet that leads to the open ocean. Because of my own experience as a windsurfing student on that same beach many years earlier, I knew what was coming next would be entertaining. Windsurfing is one of those sports—like golf and snow skiing—that is much more difficult than it looks. As I saw the kids getting in the water for their lesson, I sat down in a big wooden rocking chair on the boardwalk to watch the drama unfold.

The excited kids got on their boards and paddled out to the middle of the water behind their instructor. I couldn't hear exactly what the instructor was saying, but I could tell from his hand motions that he wanted them to pull up their sails, sail about fifty yards to a buoy, and then come back to the beach.

As the wind gently swirled and the waves began to roll, I knew what was about to happen. All of the students got up on their boards and began tugging on the ropes attached to their sails that were lying in the water. As they were fighting to pull their sails up, a few began to drift toward a marsh, others toward the boats at the dock, and two were being pulled by the tide down the inlet on the way to the open water. The instructor waved his hands and shouted instructions, but with little effect.

One girl finally pulled her sail out of the water only to have it hit her and knock her off the other side of her board. Undaunted, she quickly got back up, steadied her sail, and started moving slowly toward the buoy. Seeing this persistent young woman in action inspired a couple of the guys near her to believe that this windsurfing business must be doable after all, and they soon got their sails up and began sailing slowly in the general direction of the buoy. Three or four of the kids now inched toward the goal of buoy, but most of the others were still fighting with their sails and drifting in all directions. Two were pulled out by the tide and disappeared around the corner past the beach, where I noticed two others who still sat on the beach with their boards, hesitant to get in the water.

A few minutes later the instructor told them all to make their way back to the beach. As they came in, some were sailing, some were lying on their boards paddling, some were in the water swimming with their boards in tow, and the two who had been pulled out by the tide had to be picked up by a pontoon boat. When they all got back on the beach, something very surprising happened. Despite their limited windsurfing success, the kids were animated and even ecstatic. They jumped around, gave each other high fives, and pointed to where they had sailed or—more accurately for some—paddled. Even the two who had been picked up by the pontoon boat were talking about how they had not been defeated by the killer tide. They were all acting as if something magical had happened to them. And you know, I think it really had.

As the celebration continued on the beach, I looked at the pontoon boat parked right next to them and thought about how easy it would have been just to put all those kids in that boat, drive them around the buoy, and then bring them safely back to the beach. Then they would have been spared the strain and struggle. But I also knew that the real purpose and value of showing up on the beach was the individual battle they each went through by themselves. And ironically, it was their individual struggles that created the sense of unity and team spirit among the whole group.

My vantage point on the boardwalk that morning was not unlike the vantage point policymakers in Washington have of the human drama that plays out every day across America. Just as with those windsurfing students, there are Americans who are left behind on the beach and never get started; there are those who are in the middle trying, but failing and being dragged down by the tide; then there are the masses in the middle who are out there fighting with varying degrees of success; and finally, there are the few who are super-successful.

You would expect the political left to look at that scene and say that it's not fair, it's not equal. Everyone does not have the same opportunity to succeed because some of the kids lack the necessary capabilities to achieve the goal. Some are disadvantaged and effectively disabled by their past. To paraphrase the words of former Supreme Court Justice Thurgood Marshall, it is not equal because everyone is not getting the same thing, at the same place, at the same time.

On the political right, some might look at the whole situation and say, "What's the problem? Why change anything? Just let them struggle on their own."

Democrats often insist that the only compassionate solution is to put everyone in the same boat. Republicans could, but often don't, argue that maximizing individual liberty and freedom is the most compassionate solution. Those who can sail should be allowed to experience the joy and exhilaration of success while those who can't should be granted the dignity of being allowed to make the effort. Those who fail are likely to be helped by those who can sail, and, as a last resort, the pontoon boat is standing by in an emergency.

Yet, too often, Republicans do not make this case, and unwittingly surrender their basic principles in order to get elected. In my five years in Congress, it has become clear to me that very few politicians, even Republican politicians, are fighting for policies that promote individualism and independence. Most Republicans have essentially given up and have agreed to put everyone in the same boat, but to save face, they sometimes muster up the courage to insist that it must be a cheap boat.

I am a conservative because I support solutions that maximize individual liberty and freedom. I believe that putting people in the same boat, even if it's a cheap boat, deprives individuals of their dignity and God-given, unalienable right to liberty. Freedom is the operative force that makes America and all its institutions work. "The aspiration toward freedom is the most essentially human of all human manifestations," American philosopher Eric Hoffer said. Freedom turns hopelessness into opportunity; ideas into enterprises; and factions into communities. And by making decisions for people, we strip freedom from people and undermine our civilization.

John Adams said that the American Revolution was won in the hearts and minds of Americans before the first shot was fired. Today, America will remain strong only if we confront an enemy that is an even greater threat than terrorism, the growing dependence in the hearts and minds of Americans on "cheap boat" solutions sold in Washington. It's time for conservatives to remember what made America great and boldly fight for true freedom and individual liberty.

———————

JIM DeMINT of South Carolina is serving his first term as a U.S. senator.

P. J. O'ROURKE

· ★ ·

THE SHOCKING CONVICTIONS AND
ASTONISHING GRABBINESS OF THE LEFT

Faith and greed make me a Reagan conservative. I admit to
conventional amounts of both. But it is the shocking convictions and astonishing grabbiness of the Left that impels
me to the Right. Since the eighteenth-century beginnings
of foolish modern political thought, the utopians, anar-
chists, syndicalists, communists, socialists, progressives,
liberals, members of the Democratic Leadership Conference—
let the Jacobins call themselves what they may—have
wanted everything. They don't want to improve institutions, laws, and material well-being. That's not enough for
them. They want to improve you. Being kind to minorities
and the poor isn't sufficient; you have to *identify* with the
oppressed. Proper custodianship of natural resources isn't
adequate; you must be *sensitive* to the environment. Tolerat-
ing people whose values, tastes, and modes of existence are

peculiar or distasteful won't do; you need to *celebrate their diversity.*

The Left does not want a world of people with better lives. It wants life in a world of better people. God made people. To remake them requires all the power of God and—with the suppression of free will—then some. Thirsting for more power than God is greed on a scale to shock the most rapacious Martha Stewart.

The Left desires people to be virtuous. And, to be fair, the leftists desire to be virtuous themselves. Unfortunately, they covet virtue so much that they take moral shortcuts to achieve it. In left-wing life, as in left-wing legislation, easy words are given the credit of difficult deeds. Mere declaration of "War on Poverty" was enough to give the Great Society heroes their triumph and victory parade. When a Reagan conservative says that someone "means well," it's hardly a compliment. But when a Clinton liberal says someone "wants to do good," it's enough to excuse a Clinton (or two).

And the good that the Left says it wants to do is to be done not with their goods or their services but with ours. Taking a single thing from anybody is theft. Taking many things from lots of people is social justice—because, then, everything can be given back to everyone. By the circular logic of collectivism, the Left leads its audience around the political stage and behind an Iron Curtain.

Not that today's Left means to take power the way Ho Chi Minh and Fidel Castro did. Modern leftists are too anxious in their gluttony for command. They don't have the patience for long years of tedious organizing. And fighting the kind of revolutionary wars that led to communist dictatorships could get a caring, sharing person hurt. Why bother with all that when the bar to entry in politics is set so low? Becoming a Teddy Kennedy requires no prior accomplishments and even fewer accomplishments thereafter. Just say that man is good inside and vow to turn man inside-out. Economic greed requires buying and selling at values set by a free marketplace. Political greed can be satisfied in return for pious sentiments that cost nothing (at least to those who hold them).

Thus the Left worships government. From government all blessings flow. Government is omniscient and omnipotent, or should be, or will be soon. Some on the left say that they believe in God, as well—though I doubt the compliment is returned. The real faith of the Left is the holy goodness of humanity (as soon as the Left gets humanity reformed) and the collective cosmic solidarity of mankind. These pious sentiments have been heard since the time of that poetic genius—and political idiot—Percy Bysshe Shelley:

Man, one harmonious Soul of many a soul,
Whose nature is his own divine controul

But leave it to Lenin to put it more bluntly: "Man is god to man."

We've seen just what kind of god, with Stalin, Mao, and Pol Pot. But suppose that an example of the virtuous "New Man" imagined by the Left reigns over heaven and earth. Call him "Ben." Or call him "Jerry." Anyway, Adam sacrifices all his ribs and half his backbone so that the Garden of Eden is representative of the full spectrum of human sexuality. Endangered species go first into the Ark. (Now, how do we get those brontosaurs out of the vegetable garden?) Moses is called to the mountaintop to retrieve the Ten Thousand Commandments cajoling the Israelites to be "in touch with themselves" and deploring behavior that's "hurtful and divisive." Joshua blows his horn and the residents of Jericho join in on recorders and tambourines. There's no capital punishment in the Judea of Pontius Pilate. Jesus does three to five in a minimum security imperial pen. He writes *The Gospel of Prison Reform* and starts a socially conscious, sustainable small business by using his heavenly powers to invent refrigeration. The symbol of universal salvation is an ice cream sundae. We are blessed with an infinite number of cleverly named delicious flavors. But we are required by law to use someone else's tongue to lick them.

P. J. O'ROURKE is a correspondent for the *Atlantic Monthly*. He has been a journalist for thirty-two years and has covered news events in more than forty countries. Mr. O'Rourke is also a regular contributor to *The Weekly Standard* and *Automobile.* He is the author of ten books, including *Parliament of Whores, All the Trouble in the World, Eat the Rich,* and, most recently, *The CEO of the Sofa.* He was editor in chief of the *National Lampoon* from 1978 to 1981 and international affairs correspondent for *Rolling Stone* magazine from 1985 to 2001.

PHIL GRAMM

· ★ ·

EFFORT AND DETERMINISM

It was a lifetime ago, but I remember it like it was yester-day. Driving down the road in our old Plymouth, my mama would slow to a crawl in front of Doctor Dyke's house, the nicest place in town. She'd say to us that if we studied hard in school and worked hard in life, someday we could live in a house like that.

My mama had a simple but powerful worldview in which effort and determination prevailed over all obstacles. Maybe not every time, but often enough that the exceptions didn't count for much. Sure, it was better to be pretty, smart, and rich than to be homely, average, and poor, but in the end, being homely, average, and poor were not insurmountable obstacles or valid excuses for failure. In Mama's view, the world was fair if it afforded opportunity, and opportunity was what America was all about.

Her lessons extended well beyond a slow drive past the

doctor's home. One time she left my brother and me in a furniture store for a very long hour while she ran errands. This was a place and time when children could be left alone in safety with the recognition that friends and neighbors would not only look out for them without being asked, but report on their transgressions, too. Our instructions this day were to watch Mr. Rowe, the proprietor, to see what we might learn.

On the way home, she administered the test: What had we observed about the wealthy storekeeper, Mr. Rowe? I observed that old Mr. Rowe was a funny-looking little man. That was not the right answer, and Mama's look burned the retina in the back of my eyeballs. My older brother, Don, was the worst kind of brother you could have. He never lost his glasses, and he always knew the right answer. Well, Don reported proudly that Mr. Rowe never once sat down, he said hello to everybody who came into his store, and he worked continuously. This observation was the correct answer because it validated my mother's philosophy.

In those days she was a practical nurse who worked the 7 a.m. to 3 p.m. shift at the hospital, and then often nursed patients in their homes from 3 p.m. to 11 p.m. Her home-care patients were wealthy, at least compared to us, and she thought that the rich people she met seemed to share certain traits. They all worked harder than we did, she said, and they had "sharper tools." These "sharper tools" referred to the fact that they were better educated than us.

Fast-forward now past my failing the third, seventh, and

ninth grades. My invalid father died, leaving my mother $8,000 of GI insurance. With a tough love lecture from my mother and brother that I was running out of opportunities, my mother took what was left of the $8,000 and sent me away to a military school. I graduated with honors. After running out of money at the end of my freshman year in college, I worked a year and a half and attended night school. When I had accumulated some money, I went back to the University of Georgia.

The collage of courses I had taken by correspondence and in night school fit into no degree plan. In trying to pick a major, I happened to visit the National Science Foundation office at the university where I saw a poster that showed that in 1962 the highest paid new Ph.D.s were in economics. I didn't know what economics was, but it impressed me that obviously somebody thought it was important. I enrolled in my first economics course and by the end of the week, I was so interested that I spent the whole weekend reading the textbook.

I didn't know that people actually knew the things that I learned in economics. Economics explained the world I grew up in and by and large gave the theoretical and empirical underpinning to the way my mama saw that world. It didn't take me long to figure out that these were powerful ideas that could change the world and empower ordinary people to do extraordinary things.

Ultimately, my biggest discovery was that freedom, economic as well as political, was the only idea worth fighting for. As Pericles said in *The Funeral Oration,* "The key to happiness is freedom and the key to freedom is a brave heart." As an economist, a professor and a citizen, a congressman and a senator, I fought for freedom in all of its essentials, against all enemies foreign and domestic.

In my very first campaign speech, I said, "The American people feel a sense of helplessness. They know big government is not working, but they don't know what to do about it. What we need today is a new wave of leadership to fulfill the ideals and aspirations of a revolution which occurred almost two hundred years ago. In the coming struggle for the survival of the American experiment, I mean not only to participate, I mean to lead."

THAT WAS 1976, AND BOTH RONALD REAGAN AND I LOST OUR elections that year. But two years later I went to Congress, and four years later Ronald Reagan became president. His driving belief in freedom remade the world, but it also represented a victory for a widow and two small boys who had noticed the impact of freedom everywhere they went, including out the window of an old Plymouth on a dusty road in Georgia.

PHIL GRAMM was a U.S. senator from Texas from 1985 to 2002.

CHRISTOPHER COX

· ★ ·

CLASSIC IDEAS WORTH CONSERVING

The classical liberal ideas underlying the founding of the American government—freedom of the individual, government as servant and not master of the people—are timeless. They are worth conserving for our age and for our heirs.

Our Declaration of Independence and Constitution do not legislate prosperity, equality of wealth, or even happiness. They contemplate a political system and government institutions that free individuals to build prosperity, work for financial reward, and pursue happiness.

The difference is crucial: the Founders recognized, as we must in our own time, that the objects of any political system will never be perfectly achieved. Inequality, poverty, the arrogance of elites, environmental degradation, unfair competition, inadequate health care, and oppressive taxation all represent a status quo that Republicans seek to overthrow. Political conservatism does not imply preservation of the status quo, but rather preservation of the ideals and the republican form of government that will help us to change it.

This, too, is why I am a Republican as well as a conservative. The Republican ideal, so succinctly stated by our first presidential candidate, John C. Fremont, is freedom: free minds, free markets, free expression, and unlimited opportunity. In the nineteenth century, a political "liberal" pursued these goals—a socialist, on the other hand, sought greater government control over individual passions and perceived selfishness.

Today, conservative Republicans are the champions of free minds, free markets, free expression, and unlimited opportunity. The leading organized opposition to these ideas comes from the Democratic Party in the form of "politically correct" speech, government-funded television and radio, socialized medicine, government ownership of businesses from railroads to power plants to package delivery, heavy-handed regulation of private enterprise, and an insistence on equal outcomes instead of equal opportunity.

As the organization that led the fight to abolish slavery, enacted constitutional civil rights for all persons, and established voting rights in the Constitution for women and for people of all races and creeds, the Republican Party stands now and has always stood for liberality in the classic sense: freedom of the individual.

Inherent in the concept of freedom is its obverse, personal responsibility. Conservatives recognize that if individuals are to be truly free to choose, this freedom must encompass wrong as well as right choices. But if one's neighbors are taxed to subsidize the costs of wrong choices, as liberals insist, then the entire system of incentives for right choices and discipline against wrong choices is undermined. Making individuals accountable for their actions—their criminal conduct, their business failures, their failure to learn a trade or a profession, their poor diet—is the bulwark of everyone else's liberty.

Conservatives understand that freedom is indivisible. Political, economic, social, and religious freedom are all facets of the same gem. Modern-day liberals, on the other hand, cleave to the fiction that you can be "free" even as government dictates the use of your property, picks your doctor for you, regulates your choices in the marketplace, forces your membership in labor organizations with which you disagree, confiscates your money to pay for others' abortions and "art" that offends you, and empowers rogue lawyers to extort billions from others in your name through government-sanctioned class actions of which you are unaware and which you haven't authorized.

Conservatives recognize that only when individuals enjoy all of their freedom is the pursuit of happiness possible; only then will the society of individuals achieve its potential to create wealth, knowledge, and the betterment of the human condition.

Finally, conservatives believe the freedom bequeathed to

us in America should not be ours alone, but must become the birthright of all mankind. We understand that if "cultural condescension," as Ronald Reagan termed it, is allowed to provide a rationalization for condemning billions of men and women to live in despotism and poverty, then not only their freedom but ultimately our own will be lost.

Conservatives know that freedom isn't free, that it must be earned and protected in every generation. I deeply respect the sacrifices that our forebears have made so that every American and so many others on this earth can be free. That heritage is worth conserving for the future. That is why I am a conservative.

CHRISTOPHER COX is a U.S. representative from California and the highest-ranking Californian in the Republican leadership in Congress. He heads the new Homeland Security Committee in the House and is chairman of the House Policy Committee.

THE RT. HONOURABLE
LORD DOUGLAS HURD

· ★ ·

DIFFERENT SOILS BEARING CONSERVATISM

By definition conservatives differ from each other across the world—because conservative beliefs grow up from the different soils in which they are rooted. My own background was deeply old-fashioned, rooted in the conservatism of English villages and market towns in the 1930s, 1940s, and 1950s of the last century. I was brought up on a farm in the Wiltshire Downs; my grandfather was the local Conservative Member of Parliament, and my father followed him into the House of Commons representing the constituency next door. The Conservative Party in my boyhood seemed part of a tradition of public service. We accepted that the Labour and Liberal parties were different bits of the same tradition, and elections in our part of England were full of color and good-tempered noise. At the age of five, I was dressed in a blue coat with my two brothers and put in the front row of the village meeting addressed by the Liberal candidate in order to throw her off her stride. In the days before television, such meetings were well attended and reported almost word for word in the local press.

The fundamental belief with which I grew up was of affection and loyalty to the king, to our country, and to its traditions. Change was necessary from time to time but should always emerge from those evolving traditions.

Two notable individuals whom I served added fresh dimensions to this simple thought. Ted Heath, for whom I worked from 1966 to 1974, taught me that politics was a matter of hard work and preparation. He also revealed to me the possibilities of European unity as a remedy for wars and confusion.

Margaret Thatcher, whom I served from 1979 until her downfall in 1990, took over our country at a most dismal time. She showed us that free markets, though not perfect, were more effective than a bureaucratic state in providing the goods and services which a citizen needs. More important, she showed many of us that political courage was the best remedy for pessimism.

————————

THE RT. HONOURABLE LORD DOUGLAS HURD served as Foreign Secretary, Secretary of State for Ireland, and Home Secretary in Margaret Thatcher and John Major's administrations. Currently, Hurd writes political novels and serves as Chairman of the Advisory Committee of Hawkpoint Partners.

PAUL M. WEYRICH

· ★ ·

ETERNAL TRUTHS

My father was an immigrant with only an eighth-grade education but he was the smartest man I ever knew. He was a strong conservative, and we spent hours and hours talking about why he chose that path when he came to America. He started out as a Democrat but changed parties when he felt that party no longer represented his philosophy. So at first I would have to say I became a conservative out of my admiration for him.

But as I grew older, I began to examine my own thinking. The utopian ideas of liberalism had a certain attractiveness. I certainly had many challenges to my views at the University of Wisconsin. The more I studied and the more I experienced life, the more I became convinced that my father was, in fact, correct. Traditional values are functional values. Conservatism works. It works because it is rooted in reality. Utopianism may be fine for university classrooms, but in the real world, the eternal truths, based on biblical concepts, are the only ideas that make sense. To the extent we deviate from these eternal truths is the extent to which society becomes dysfunctional. I am a conservative because I want to live in a society that works well and that provides the greatest opportunity for everyone to use their God-given talent.

PAUL M. WEYRICH is chairman and CEO of the Free Congress Research and Education Foundation, a conservative think tank.

GROVER NORQUIST

· ★ ·

AMERICA IS FREEDOM

I am an American.

America is not a race. It is not a religion. It isn't even a place. If we all moved to a distant planet, we would still be Americans.

America is freedom. Liberty.

Americans are united by the Constitution and a belief that individual liberty is the highest—indeed only—political goal. Individuals may choose to exercise their freedom in very different ways. Some create wealth. Others focus on raising a family. Or practicing their faith. Or a little of each. The sole legitimate function of the government is to create and protect that liberty. That is why we have courts, police, and a national army. To keep out, stop, or punish those who would infringe on our liberty.

And in America the conservative movement is the defender of America as liberty. The modern Reagan/Bush center-right coalition is a coalition of groups and individuals that—on the issue that moves them to vote—wish to be left alone by the central government. Gun owners do not want their Second Amendment rights infringed upon. Property owners do not want their property rights interfered with. They do not want to hear that it rained last night and Al Gore owns their backyard. Businessmen and -women do not want their businesses taxed and regulated. Homeschoolers want to raise their own children. And all communities of faith—evangelical Protestants, conservative Catholics, orthodox Jews, Muslims, Mormons—all want to be able to practice their faith free of government coercion. Each part of the center-right coalition may want freedom for a different reason—but they all want liberty and understand that their liberty is dependent on creating a political movement that guarantees all its members liberty in the zone of greatest importance to them.

I am a conservative activist because I am an American. If I lived in Belgium, it wouldn't matter if I worked really hard and Belgium got 100 percent freedom. It would eventually be eaten by the French or the Germans. But when America gets it right, the world benefits. We are the city on a hill: large enough to dominate the planet through our movies, television, radio, and example of liberty that works. If liberty was lost in America, it would be lost to the world. When liberty is strong in America, it creates more liberty in faraway places where men and women rise up to demand their rights as individuals.

For our constitution declares that all men and women have the God-given right to liberty. What we protect in America is deserved by all the people on the planet. Except maybe the French.

———————

GROVER NORQUIST is president of Americans for Tax Reform, a coalition of taxpayer groups, individuals, and businesses opposed to higher taxes at the federal, state, and local levels. He serves on the Board of Directors of the National Rifle Association of America and the American Conservative Union.

FRANK FAHRENKOPF

· ★ ·

THE MAKING OF A CONSERVATIVE

When I arrived at Boalt Hall School of Law at the University California-Berkeley in September of 1962, I was a fairly moderate Republican whose father, an automobile mechanic, had been a lifelong Democrat. While at Berkeley I witnessed firsthand the so-called Free Speech Movement (FSM), which was in reality the birth of campus unrest across the country in the early 1960s.

I knew Mario Savio (the FSM leader) fairly well, as he sold bagels outside of the law school and he often came into the student lounge to debate, with some of us, his extreme left-wing positions. Any student or student organization that wanted to collect money, distribute pamphlets, and so on could do so at Cal by setting up tables or other facilities in a designated area adjacent to Sproul Plaza—the main entrance to the campus.

Savio and his cohorts, however, wanted to set up a table

to raise money in support of the Viet Cong directly in the center of Sproul Plaza rather than in the area designated for all student organizations. The university's prohibition of the FSM table in the middle of the plaza was the alleged violation of their "free speech rights."

I was personally offended by the Savio gang's tactics and was present when singer Joan Baez led the mob into the university's administration building, Sproul Hall, which they devastated in their overnight stay.

Over the next few weeks, the student strike and other actions of the so-called Free Speech Movement people convinced me of their philosophical hypocrisy. Their view of free speech was solely determined by whether they agreed with what you were saying.

I was present at a speech by former vice president Hubert Humphrey that was rudely disrupted by Savio and his gang because they disagreed with the content of Humphrey's speech. The vice president was not accorded "free speech rights" because his speech did not agree with the FSM viewpoint.

In 1964, Barry Goldwater won the Republican nomination for the presidency in the Cow Palace in San Francisco, across the bay from Berkeley, and I first became aware of Ronald Reagan, in a political sense, as he campaigned and spoke on behalf of Goldwater.

Following Goldwater's defeat, I was profoundly influenced by Reagan's regular radio talks on various subjects of importance to the American people. The Berkeley experience, the Goldwater nomination, and my exposure to Ronald Reagan completed my transformation from moderate to conservative.

———————

FRANK FAHRENKOPF is former chairman of the Republican National Committee, serving during six of Ronald Reagan's eight years in the White House. He currently serves as president and CEO of the American Gaming Association. He is also cochairman of the Commission on Presidential Debates.

CRAIG THOMAS

· ★ ·

THE CONSTITUTION'S GROUNDWORK

Many of the decisions made in the governance of our country and in our local communities are frequently described as conservative or liberal. To me, conservatism means developing a framework within which individual citizens have the opportunity to make decisions for themselves, and that the function of government is to protect people's rights and provide the opportunity for individual opportunities. I believe the Constitution of our country laid the groundwork for that kind of government, and of course the success of the Constitution has been proven. The politics of decision making often require that conservatives look as if they are not doing everything for everyone. Liberals, of course, take political advantage of that perception. I think the bottom line is that the role of government ought to be limited in size, oriented toward individual freedom, and designed to assist people in living their lives and developing their success, without excessive involvement or dependence on government. I am very proud to be a conservative and to promote these concepts.

CRAIG THOMAS is a senator from Wyoming. He is the Senate's Republican chairman on the Congressional Oil and Gas Forum and a member of the Senate Finance Committee, where he chairs the International Trade Subcommittee. He also serves as cochair of the Senate Rural Health Caucus.

JAY NORDLINGER

· ★ ·

I'M A REAGANITE

First, to dispense with the question of what a conservative is. Or not to dispense with it. You can get bogged down for days debating the meaning of "conservative." I always skirt this debate by saying, "I'm a Reaganite." In fact, that's about the best thing Ronald Reagan ever did for me: give me something to call myself. Instead of delivering a lecture on the Scottish Enlightenment—or on the peculiar evolution of the terms "conservative" and "liberal"—I just say, "Hey, I'm a Reaganite." Some people think Reagan was a mainstream conservative, a right-winger, a progressive conservative, a genuine liberal, a Neanderthal—whatever. But at least people have a sense of what Reagan stood for.

I was not born with a conservative, or a Reaganite, spoon in my mouth, that's for sure. I was brought up in a left-wing environment, and there's probably more than a little backlash in me. Early on, I learned to "question authority": to question the prejudices and assumptions around me. In my environment, "conservative" meant bigot, ignoramus, war-monger, Darwinian, scorner of the poor, and so on. But I soon outgrew this stuff. And a lot of it was personal. I wondered, "If the Left is the party of love and compassion, how come so many of them are such . . . ?" Well, I'll leave the expletive deleted. The "liberals" of my youth, in Ann Arbor, Michigan, were a decidedly illiberal bunch: close-minded, dogmatic, intolerant of dissent. I could not subscribe to their orthodoxy. Plus, I was, as now, very religious, and "liberals" didn't think much of people like me. Relatedly, they had a disdain for ordinary people. I was particularly appalled at their view of black Americans: It was condescending, at best.

I should also say that I was an anticommunist, and I thought that people who loved humanity should at least oppose those governments that killed humanity en masse: in China, in Cambodia, in the Soviet Union, and so on. How could lovers of humanity adorn their walls with posters of Mao and Guevara?

The experience of the Reagan years wiped out any leftism I might have had. I saw what a program of economic growth—including tax cuts—could do. I saw what standing up to Big Labor (e.g., the air-traffic controllers' union) could do. I thought that he was spot-on about the Soviet Union, and about resistance movements in places like Nicaragua. And I settled on the pro-life position: I simply could not accept that, in aborting a child, a woman was merely exercising her sovereign right over her own body. For there was another body, another life, involved.

I once asked a colleague of mine—a woman in middle age—why she was a conservative. She answered, "Because not having an abortion was the most important thing I ever did."

I should also say that I was greatly influenced by the sociologist Charles Murray and his researches into the welfare state: he wasn't interested in saving the federal government any money; he wasn't a fiscal conservative; he simply cared about the effect of welfare on its intended beneficiaries— and he said it was no good for them, and I believed him.

No, they don't make liberals like they used to. Thomas Jefferson would have nothing in common with today's liberals. For what has liberalism come to mean? Speech codes, race preferences, and abortion on demand. Great. Just great.

I will attempt a little credo—nothing fancy, just straight-out. Why am I a conservative? Because I value the individual, and individual responsibility. Because I know the power of a free economy. Because I'm pro-life. Because I'm for colorblindness, and an end to the racial spoils system, and an end to race-baiting. Because I believe in the old *E pluribus unum*—Out of many, one. Because I detest the stranglehold of the teachers' unions over public education. Because I think we should stand up to bullies abroad. Because I think that military preparedness is essential. Because

I think that the courts should remain courts, instead of false legislatures. Because I think we should be for sensible environmental conservation, instead of absurd Earth worship. Because I think that Western civilization is worth preserving, from the Old Testament to (at least) Stravinsky.

And basically because I don't stay up nights, tossing and turning, worrying about how to control other people's lives. I have enough problems with my own. I think that people should work out their own salvation with diligence. Someone once said—was it Samuel Gridley Howe?—that we should all walk the path alone, but look back, every now and then, to be sure that others aren't lagging too far behind. And yet, government does not have a monopoly on charity and charitableness. Far from it.

Anyway, that's why I'm a conservative. A Reaganite. Whatever.

JAY NORDLINGER is managing editor of the *National Review* and a reporter, essayist, and critic.

ROBERT LIVINGSTON

· ★ ·

THE MARKS OF A CONSERVATIVE

A modern-day conservative is in fact a liberal in the sense that our forefathers intended when they drafted the Declaration of Independence and our Constitution. I consider myself to be one and the same with both terms.

As a person who believes that this country provides opportunity for all who may wish to avail themselves of its bounty and freedom to engage in the pursuit of happiness, I am most definitely a conservative.

As one who believes that with freedom comes responsibility, self-discipline, and the respect for one's neighbors, I am a conservative.

The United States is a great and wonderful nation that unites peoples from all corners of the globe to join and participate in a free and democratic way of life, and I believe so strongly in her promise that I would shed my life and all I hold dear to protect her against threats from tyrants, dictators, and oppressive governments of all kinds to ensure her safety. This, I believe, is the mark of a conservative.

To live within one's means as a person or as a people is most certainly a trait of conservatives. To resist the dictates of an intrusive, profligate, or burdensome government is conservative as well.

To appreciate the sacrifices of our forebears, to hold in one's heart the significance of their accomplishments that have led to the grandest of all experiments in social history, the creation of one nation under God, free and just, which in turn has set in motion the opportunity for more people to live in freedom today than at any time in world history, is endemic to the term "conservative."

To be proud to be an American—that is most conservative.

To join with my wife, Bonnie, and raise our four children in such fashion that in their own ways, they are each proud conservatives as well, then, I am a fulfilled conservative.

ROBERT LIVINGSTON is a retired member of Congress and the founding member of the Livingston Group, a government relations firm in Washington, D.C.

SUZANNE FIELDS, Ph.D.

· ★ ·

FROM CULTURAL LIBERAL TO
CULTURAL CONSERVATIVE

I was the mess sergeant for Norman Mailer's *Armies of the Night.* I threw "the liberal party" described by Mailer in his Pulitzer Prize–winning book. These were the troops, brimming with rage, fire, and zeal, intoxicated with virtue (and other stuff), who had come to Washington to close the Pentagon, to spike the engines that ran the war in Vietnam.

My guests were my heroes. I was a groupie for the literary glitteries, feeding them their last hearty meal before they left to make the good fight against the cops, whom they devoutly hoped would arrest them. They didn't quite meet my heroic expectations. Norman Mailer got drunk telling Robert Lowell, the pacifist poet, that he would rather be a WASP poet than a Jewish novelist. Robert Lowell got drunk listening to him and mumbled gibberish that sounded something like poetry. Dwight Macdonald, an essayist for *Esquire,* was the social butterfly of the evening, displaying a button on his lapel with a tiny photograph of Rosa Luxemburg, the socialist revolutionary murdered in Berlin in 1919. He would only talk to guests who could identify her.

Paul Goodman, author of *Growing Up Absurd,* the student bible demanding that education be made "relevant" and all grades be abolished, fell asleep on my living room floor. We didn't know then that the title of his book would become the verdict on the decade of the 1960s.

That party started moving me from left to right, I realized later, from cultural liberal to cultural conservative. The buffoonery gave me the insight to see through nonsense passing for intellect. My husband and I were raising our family, and I began to see the goofiness of many of the notions my liberal friends and I embraced. Education was dumbed down and we cared more about nurturing "self-esteem" than cultivating academic rigor. The sexual revolution had gone too far and our children were vulnerable. This was an era when a conservative was a liberal with a daughter in junior high school. I became concerned with "family values" before the phrase became a cliché.

Ronald Reagan revived my patriotism. He brought back an appreciation for the sensibilities of my parents' generation, which I had proudly rejected. I visited the Soviet Union before and after glasnost and saw the reality behind what was truly an evil empire. My father, who was born in Russia, once remarked to me that his father wasn't the smartest man in the world but he was smart enough not to miss the boat to America. I was a conservative when I finally understood what he meant. I wanted to conserve those ideals of freedom, even if sometimes honored in the breach that welcomed my grandfather to these shores.

SUZANNE FIELDS, PH.D., is a syndicated columnist in more than thirty newspapers and the author of two books, *Like Father, Like Daughter: How Father Shapes the Woman His Daughter Becomes* and *How the Cookie Crumbles,* a collection of her columns.

EDWIN J. FEULNER, Ph.D.

· ★ ·

WHAT CONSERVATISM IS NOT

Let me define why I am a conservative by saying, first of all, what conservatism is not. Conservatism is not a political party, with a shared platform, hashed out by lobbying and majority vote. According to Russell Kirk, it is "neither a religion nor an ideology," and it possesses "no Holy Writ and no Das Capital to provide dogmata." It is, instead, a broad social movement of diverse but reinforcing beliefs, gathering travelers on the same journey—pilgrims who argue over the topography of their promised land but who move in the same direction.

This social movement is built on a foundation of a consistent philosophy. I am referring to the tide of ideas running through our times, lapping on the shores of many nations— the intellectual triumph of the philosophy of freedom over the various utopias of central planning. A philosophy that believes, that while government is necessary, it should be limited, and that government is best when it is closest to the people.

Conservatives tend to believe there is a close and necessary connection between property and freedom—that economic freedom is an essential part of human freedom. Economic markets, when left to themselves, often have unexpected and positive social benefits. And this should not surprise us, because they are based on cooperation rather than coercion. It has been noted by conservatives that to prosper as a socialist you need to threaten people, while to prosper as a capitalist you need to please them.

Conservatives teach that the order of society depends directly on the moral order found in the souls of citizens—that freedom must be tempered by internal restraint, so our laws can be permissive while our society is not.

Conservatives believe that individual liberty is protected by the preservation of national sovereignty, making national defense a high moral duty.

Conservatives judge social policies by their outcomes not their intentions, arguing that humanitarianism should do something positive for actual humans.

And, though this is not universal, most conservatives share a sense of reverence, a belief in two worlds—one physical and one moral and spiritual—that stand in judgment of our own.

These are some of the themes of a conservative philosophy. I find all of them compelling and none of them contradictory. The danger of division, as I have seen it, comes when one conservative value—either freedom or moral order—is elevated to an absolute, to which everything else is sacrificed. So perhaps the highest conservative virtue is prudence—the right balance of valid and competing truths.

So, taken together, this set of beliefs underlies my conservative philosophy and answers the question of why I am a conservative.

EDWIN J. FEULNER, PH.D., is president of The Heritage Foundation. A winner of the Presidential Citizens Medal, Dr. Feulner has authored five books and has a biweekly syndicated column that appears in more than five hundred newspapers.

DONALD J. DEVINE, Ph.D.

· ★ ·

ALL OTHER POLITICAL PHILOSOPHIES
HAVE FAILED

I am a conservative because all other political philosophies have failed. And the poor, modern world badly needs the vision of a positive future that only conservatism can provide.

As recently as the 1950s, liberals could claim "the end of ideology," for there was no conservative or even leftist alternative to establishment liberalism. Progressive Democrats controlled everything. While communism had international support and the power of the Soviet Union to enforce it, social democratic welfare statism was supreme domestically in America and Europe. While liberalism labeled itself democratic, its mode of governance required popular deference to expert, best-and-brightest bureaucrats and judges who were to make all of the important decisions—based upon a supposedly objective, scientific, and legal authority that could not be questioned by ordinary Americans who did not have the necessary understanding.

There were traditionalists and libertarians who opposed that dominant liberal ideology, and there were Republicans who were "do it slower-than-the Democrats," moderates. But there were no conservatives in the modern sense. Modern conservatism was invented at *National Review* magazine in the midfifties, primarily by editors William F. Buckley Jr. and Frank Meyer, who promoted a vision that they saw as an alternative to the existing suffocating, bureaucratic welfare state. Buckley labeled this as "up from liberalism." As befitting conservatism's positive view of common sense and tradition, the new doctrine was not planned but grew from the interactions of its creative but divided staff, which needed some common ground from which to publish a coherent enterprise. Meyer dubbed the intellectual product "fusionist" conservatism.

Fusionist conservatism's highest value was liberty, but it was a freedom to be used responsibly as a means to pursue traditionally defined and virtuous ends. The conservative alternative was based on freedom: free citizens, free markets, voluntary associations, local governments, unfettered businesses—especially small businesses—and capitalism generally. But freedom was not the end. Judeo-Christian morality, the family, religion, local communities, and national patriotism were the values Meyer defined as uniquely Western that both supported freedom and made it meaningful. The formula was to utilize libertarian means to pursue traditional ends, uniting the previously divided separate political strands of the Right. This modern conservatism appealed to the traditional American desire for both freedom and community responsibility, so well expressed by its greatest early chronicler, Alexis de Tocqueville. Indeed, modern conservatism was an attempt to revive Tocquevillean citizen optimism, a spirit that had become so frayed in modern America that President Jimmy Carter could declare a future of limits, not of opportunities. The best and brightest insisted that economic stagnation was the necessary result of the monetary policy required to support the popular demand for a large and growing welfare state. Freedom undirected by government experts was passé, out of it, not very bright.

Barry Goldwater first awakened a new spirit of freedom politically but it took the optimism and decency of Ronald Reagan, who also was an early devotee of *National Review,* to provide the inspired leadership that produced the mature conservative movement. This formula inspired millions of activists, additional conservative journals, new think tanks, and political action organizations, which ultimately controlled the Republican Party and allowed Mr. Reagan to be nominated. But it took double-digit inflation, soaring interest rates, and economic stagnation to prove the bright progressive elites in government had failed, which finally ushered in the conservative era. President Reagan's successes in limiting the welfare state, the fall of the Berlin Wall and communism, and, ultimately, the 1994 majority in the House of Representatives confirmed the victory.

Once the conservatives won power, however, they found the job of reforming the welfare state overwhelming and began acceding to it. While Bill Clinton was proclaiming the era of big government over, Republicans in Congress began increasing domestic government spending. Not that spending was itself the goal. As President Reagan had put it, in his most Tocquevillaen manner, "We're not cutting the budget simply for the sake of sounder financial management. This is only the first step in returning power to the states and communities, only a first step toward reordering the relationship between citizens and government." In fact, he did reduce nondefense spending from 17.9 to 16.4 percent of GDP, unleashing an enormous prosperity. When George W. Bush won the presidency, he proclaimed a 4 percent growth in discretionary domestic spending was the maximum responsible increase. Before 9/11, he had agreed to a 6 percent increase, the Clinton average. By the end of his first Congress, spending had actually increased 9 percent. Only nineteen House Republicans voted against a new drug benefit for Medicare that would add $400 billion in spending over ten years and a total of $7 trillion in future obligations.

By the end of President Bush's first term, many conservatives had accommodated to the liberal welfare state philosophy that conservatism had been formed to overthrow. As important as the tax cuts were, they could not overcome the massive new spending. Like the earlier Democrats, they could not solve the big economic problems before them. Facing a shortfall in funding of an incredible $36 trillion (not billion) for Medicare, plus $7 trillion for Social Security and $3.5 trillion on the official books, the response of the conservative party in Congress and the White House was to add an additional $7 trillion burden toward national bankruptcy. Since more nations fall from bankruptcy than even war, this can be safely labeled a second malaise, again of Carteresque proportions, but this time a malaise of the Right—a fear that runaway spending was required to keep the masses politically contented.

In the 1950s, conservatism broke with both progressive democracy and conventional republicanism because both had accepted the welfare state and had no answers about how to confront its central problems, which were leading to national decline. The Reagan revitalization arrested that decline, but bankruptcy, hyperinflation, and unrestricted government spending once again haunt the political landscape as the baby boomers begin retiring in a mere ten years. The only possible solution is a return to the original vision of conservatism, one based on overcoming the welfare state rather than participating in its decline—or Mr. Carter may yet prove to be the superior prophet about America's future.

DONALD J. DEVINE, PH.D., is vice chairman of the American Conservative Union. He is also the Grewcock Professor of American Values at Bellevue University, a *Washington Times* columnist, a political and management consultant, and an adjunct scholar at The Heritage Foundation.

PETER BRIMELOW

· ★ ·

HOW CAN MAN DIE BETTER?

I recently learned with delight that the Indian Army has inscribed, on the monument erected in Tawang Province to its soldiers who fell in the 1962 border war with China, the famous lines from Lord Macaulay's "Lays of Ancient Rome":

And how can man die better
Than facing fearful odds
For the ashes of his fathers
And the temples of his gods?*

Of course, it's entertaining to see that the robust attitudes of the Raj, Britain's Victorian empire in India, are still going strong more than half a century after India became

*Macaulay quote: http://www.bartleby.com/100/405.38.html; full poem: http://www.jerrypournelle.com/reports/jerryp/lays.html#Horatius

independent. And the conservatism of India's formidable army, which includes regiments that can trace their lineage back to the Napoleonic War, has been a bedrock factor in the country's survival into the postimperial age.

But beyond that, it's always seemed to me that Macaulay's verse is a litmus test of conservatism. It either speaks to you or it doesn't. And it spoke to me when I first read it as a boy in the north of England—ironically in a story about a British garrison under siege during the Indian Mutiny, in the yellowing pages of a boy's magazine that had somehow been saved since my grandfather's childhood, two world wars and various social revolutions earlier.

Macaulay himself was actually a liberal—admittedly a nineteenth-century liberal, so arguably he might have been what today is called a libertarian. But I'd have to say he was probably also something of a twentieth-century liberal, in his iconoclasm and his brash rationalism. Nevertheless, here he chose to assume the voice of an imaginary conservative Roman ("much given," he quipped in his introductory essay, "to pining after good old times which had never really existed." Sound familiar?). The result is a moment of true artistic genius.

It's vital to note that this verse is not vainglorious and chest-thumping, but somber and stoic. The previous lines are:

Then up spake brave Horatius
The Captain of the Gate:

"To every man upon this earth
Death cometh soon or late . . ."

In other words, Horatius's point is not that death facing fearful odds is so wonderful—but that there is no wonderful alternative. So why not?

Note also that Macaulay has Horatio fight for "ashes" and "temples"—that is, not in the hope of saving any living family members, or in the service of any gods in whom he has confident faith. The implication: he could have no family members and no hope of divine providence—but his death in battle would still have symbolic resonance.

The core of conservatism, it seems to me, is this recognition and acceptance of the elemental emotions. Conservatism understands that it is futile to debate the feelings of the mother for her child—or such human instincts as the bonds of tribe, nation, even race. Of course, all are painfully vulnerable to deconstruction by rationalistic intellectuals—but not, ultimately, to destruction. These commitments are Jungian rather than Freudian, not irrational but arational—beyond the reach of reason.

Leftists often say that conservatives are motivated by "hate," because its recognizing these loyalties tacitly implies, by definition, that these loyalties have limits. But the truth is that conservatism is motivated by love.

In my own case, growing up in Britain in (ahem!) the 1950s, I felt keenly and bitterly the humiliations heaped upon the dying empire, notably the aborted Suez expedition. And in the 1960s, I immediately and instinctively identified with the Americans in Vietnam, despite following the war through the very dark glass of the British media, including our family newspaper, the very liberal *Manchester Guardian* (now removed to London as the *Guardian,* and even further to the left).

Ironically, now that an Anglo-American empire is being reestablished in the Middle East, I find myself more cautious than many of my old friends in the American conservative movement. This is certainly because of those early memories of defeat. But my belief in the centrality of the national interest—the American national interest, of which I regard Britain as a subset—does not waver.

We come now to the great switch. I have no doubt that most contributors to this book are going on about capitalism, the free market, and even liberty.

And I agree that these are natural outgrowths of conservatism. But they are outgrowths—epiphenomena of the post-Enlightenment modern age. Conservatism itself is premodern.

Why has capitalism proved so compatible with conservatism? I think the answer lies in the term that Australians use for capitalism: "economic rationalism." (Diversity is not strength, despite advertisement, but we are fortunate that the English language has so many growth points.)

Capitalism *is* a rational system. It can only function in a sophisticated political culture. The Austrian economist Freidrich von Hayek famously observed that the appeal of socialism is perennial because human beings have spent the overwhelming bulk of their history in hunter-gatherer bands and naturally think in terms of face-to-face relationships. When rents go up, people find it easy to blame the evil landlord and to vote for rent control. They find it very hard to think abstractly about market forces and the unintended consequences of government intervention.

But conservatives can, because they are temperamentally prepared for "thinking about the unthinkable"—to use the title of Herman Kahn's once-notorious application of realpolitik to nuclear war. Paradoxically, this is precisely because conservatives know the limits of reason. They accept that their values are arational (see above) and don't expect to derive them from a rational process. So they are prepared to reason dispassionately.

In contrast, leftists do believe they derive their values from reason. Consequently, they are always getting into a state of hysteria if reason threatens to take them in some direction that they don't, in their unexamined way, want to go.

This explains one of the characteristic sights of American debate: liberals in a moralistic snit. Liberals are always in a moralistic snit because their standard operating procedure makes no clear distinction between reason and morality. Everything they do and say is suffused by emotion, in an unstable and unpredictable way. Hence "political correctness" and the extraordinary range of American taboos familiar to all working journalists—for example, about race and gender. In themselves, these taboos give the lie to the idea that American political culture is as rational as advertised.

This contingent nature of capitalism is proved by the interesting fact that, in every English-speaking country, Left and Right have changed sides on it since the nineteenth century. Liberals like Macaulay supported the free market—economic rationalism—because they saw it as a way to break down the established order. Now, liberals generally oppose capitalism because they see it as a way to break down the established order. Conservatives oppose liberals, consistently.

What now? The American Conservative Movement, which I regard as the flower of democracy and the savior of the free world, to which I immigrated and in whose closing years I was proud to play a small part as a journalist, is finished. Greatly to my surprise, those who said the end of the cold war would prove fatal to it were right.

It turns out that many who joined the anticommunist coalition also harbored messianic fantasies about "global democracy" and America as the first "universal nation" (i.e., polity; nation-states must have a specific ethnic core).

I can't even be sure that those of us with the deep-structure personality type historically called "conservative" will be able to hold onto the word. Maybe we'll agree to adopt the suffix "paleo" (-conservative and -libertarian).

But we won't go away.

And that's an essay for another book.

PETER BRIMELOW is the editor of www.vdare.com and author of *Alien Nation: Common Sense About America's Immigration Disaster.*

KEN MEHLMAN

· ★ ·

I BELIEVE IN FREEDOM

I am a conservative because I believe in freedom, and because like other conservatives I understand that an adherence to its mandate ensures the greatness of our nation and the security of our world.

At home, freedom brings our nation to greatness through the politics of aspiration, or as one Lincoln scholar wrote, "the right to rise." Conservatives believe that ordinary citizens can go as far and high as their dreams will take them. And we understand that a nation's power is directly proportional to the empowerment of its citizens. So we champion individual rights, local control, and limited government, and on any number of issues—ranging from economic growth to health care to education—we know that free people will make the best decisions for themselves, their families, and their communities.

We also understand that freedom is the foundation of international peace, and that only the renewed commitment of each successive generation prevents it from collapse. Ronald Reagan, Margaret Thatcher, and John Paul II understood this inherently, and they promoted peace by their moral, military, and intellectual campaign against international communism. By speaking truth to the power of the evil empire, supporting dissidents within the Soviet Union and Eastern Europe, and expanding our defenses, these conservative leaders brought down the Berlin Wall without firing a shot.

By contrast, Neville Chamberlin and other European leaders in the 1930s provoked both World War II and the Holocaust by choosing the false peace of tyranny even when confronted with Hitler's successive aggressions—his remilitarization of the Rhineland, his invasion of Czechoslovakia, his development of dangerous weapons—they remained idle. As Churchill pointed out, these appeasers chose peace over honor, and ended up with neither. Had they acted, *Mein Kampf* would have been a collection of bad ideas, not the road map to the bloodiest and most deadly engagement in history.

Shortly after World War II, Russell Kirk wrote that conservatives inherit from Edmund Burke the ability to re-express our "convictions to fit the time." And in the twenty-first century, we again face a threat, and a choice. Do we take the path of Reagan and Churchill to confront international aggressors, stop dictators who threaten their neighbors, and bring terrorists to justice wherever they gather? Or do we adopt the false hope of peace through appeasement, only to leave our children and future generations with a much larger and more dangerous threat? Conservatives understand the mandate of freedom, and that it calls us to the path of resilience.

This calling has never been more urgent than now, in the presidency of George W. Bush. Under his leadership we rise to write the next chapter in freedom's story, one whose peaceful ending depends—as always—on our renewed commitment.

KEN MEHLMAN, campaign manager for the Bush-Cheney 2004 presidential campaign, is the chairman of the Republican National Committee.

CHUCK HAGEL

A PHILOSOPHY OF GOVERNANCE

Individuals who are privileged to hold high office should be anchored with a philosophy of governance. How do you govern? What do you believe? What is the role of government? These are fundamental questions that are part of a political philosophy.

I believe government is important and the power of government should be used carefully and wisely. Individual rights, limited government, free trade, fiscal responsibility, balanced budget, and limited foreign entanglements make up my philosophy of governance.

Government should be used to help those who cannot help themselves. Government can and has done much good for people. But it is not the answer to every challenge and injustice. Government must protect the rights and freedoms for all individuals with a consistent constitutional application to all issues.

For these reasons I am a conservative.

———————

CHUCK HAGEL is Nebraska's senior U.S. senator and serves on Foreign Relations, Banking, Housing and Urban Affairs, and the Select Committee on Intelligence. He also serves as chairman of two subcommittees—the Senate Foreign Relations International Economic Policy, Export and Trade Promotion Subcommittee, and the Senate Banking International Trade and Finance Subcommittee.

KATHERINE HARRIS

LIMITED GOVERNMENT AND
UNLIMITED PROSPERITY

I am a conservative because I believe in the fundamental values of limited government and unlimited prosperity through lower taxes, less regulation, and responsible stewardship of the public purse. Our nation's adherence to these basic principles will continue to provide the foundation for our greatness as a nation.

From the days of the Revolution, we have entrusted our dreams and aspirations not to monarchs, potentates, or the monolithic institutions of government, but instead to the creativity, resourcefulness, and determination of every individual. Liberals who propose a new government program as the solution for every problem have forgotten that the genius of the human spirit, not the machinations of bureaucrats, produced the unprecedented standard of living that Americans enjoy today.

The Great American Experiment has succeeded because

of our ability to adapt to a changing world, while remaining faithful to the enduring values of liberty, opportunity, and the unparalleled dignity of every man, woman, and child. Conservatives uniquely understand and apply this fundamental truth.

CONGRESSWOMAN KATHERINE HARRIS is the U.S. representative for the Thirteenth District of Florida. She serves on the House Financial Services Committee and the House International Relations Committee.

ORRIN G. HATCH

· ★ ·

RESTORING ESSENTIAL VALUES THROUGH THE REAGAN REVOLUTION

I haven't always been a Reagan conservative. In fact, I used to be a union card–carrying Democrat. I grew up working with metal lathers, and they drilled into me the Democratic mantra. My family was poor, and I thought anyone who was poorer than we were should have everything given to them.

Over time, I saw the difference in the lives of people who counted on handouts to keep them going, and those who struggled to stay independent and maintain their self-respect. It gradually dawned on me that Democrats were indeed compassionate, but not to the poor. They were compassionate to the union leaders who would help them stay in power. I discovered I didn't share the Democratic core values—central control and less personal responsibility—and I saw how an intrusive government could harm individuals and businesses.

I decided to run for office because I believed the country

was headed in the wrong direction. We were struggling with double-digit inflation, high interest rates, and growing unemployment. Our military strength was eroding. The nation was still reeling from the war in Vietnam and the scandals that led to the resignation of President Richard Nixon.

Congress was in the clutches of the Democratic Party, and what Congress did best was spend money. The federal budget was expanding rapidly, expenditures were far outpacing revenues, and yet no one seemed concerned or even aware that they were saddling future generations with a massive debt.

Instead, Washington was governed by the belief that government was the answer to every question and the solution to every problem. All that was needed to establish a perfect society was the right number of laws, the right number of rules and regulations, and scores and scores of federal regulators charged with making sure that everything ran smoothly. Most alarming, the practice of religion was treated with increasing hostility. The Supreme Court banned school prayer, legalized abortion, and devalued personal responsibility.

I was convinced that someone needed to stand against these trends. Someone needed to point out the deterioration of our moral fiber, the proliferation and increasing acceptance of drugs and crime, the expansion of the welfare state. There was a need to refocus attention to the diminishment of our military, the federal takeover of our local schools, and the bartering of our children's future by politicians who seemed more concerned with their next election than the need for fiscal responsibility. It was time for a different philosophy, a different kind of politician.

When I decided to run and fight for conservative values, I discovered I wasn't alone. On the Pacific Coast, a governor in California was leading a gathering conservative revolution that would soon invade the Potomac.

We believed the answer lay in lower taxes, less government, fewer regulations, less centralized power, and a wiser use of the power that must be exercised on behalf of the people. The Reagan Revolution restored the essential American values in all of us—the values of individualism and enterprise, initiative and optimism, charity and sacrifice. We had a core belief that government could lead society, but not build society. Our policies recognized that government's most important economic role was to foster American innovation and industry.

In truth, Reagan conservatism demonstrates true compassion to those struggling in America. Not the Democratic style of compassion shown through government handouts, but in creating an environment that inspires communities and individuals to take advantage of opportunities to improve their condition. They have the power to change, and government can help them lead better lives.

Our country was founded on the principle of "We the

People." As a Reagan conservative, I have a greater faith in the collective wisdom of the public—we can trust them to make the right decisions by themselves.

———————

SENATOR ORRIN G. HATCH has represented Utah for five terms.

MONA CHAREN

· ★ ·

INCULCATING NOBILITY

At the heart of the liberal worldview is the Rousseauian notion that people are basically good—it is merely institutions and corrupt civilizations that lead them into evil. I
don't know how anyone who has ever spent time on a play-
ground can believe this. While I would not go so far as to
say that human beings are innately evil, it is fair to notice
that without adult supervision and intervention, children
are capable of quite staggering cruelty, deception, treachery,
and even torture. They are also capable, thank God, of altru-
ism, sympathy, courage, and honor. If they weren't, we
might as well chuck the entire human project.

But the conservative tends to believe that while human
beings are capable of nobility, it does not come naturally,
but must be carefully inculcated. In other words, we rejoice
in and depend upon tradition and civilization to make life
livable. So when liberal educators, for example, tell us that

children are "natural learners" and that the job of the teacher is to stand aside and let each child's knowledge blossom like a wildflower, we snort. When they suggest that handing people welfare checks will not have a corrupting influence on the recipients' pursuit of work, we roll our eyes. When they assert, over and over again, that the proper approach to foreign policy is one based on diplomacy alone (never force, and never even diplomacy backed up by force), we simmer.

Conservatives, it seems to me, are far more realistic about human nature than liberals are. This is the genius of America's Founders, who distrusted power in all its forms, whether wielded by a despot or a mob. By designing a government perfectly suited to restraining our worst impulses, they created a system that permitted the "better angels of our nature" to thrive.

There does seem to be a limit to liberals' starry-eyed view of human beings. When it comes to the United States, their naive belief in people's essential virtue seems to evaporate. They attribute the worst possible motives to each and every American action around the world, and they tend to believe that our domestic policies are sadly inferior to those of most other industrialized nations. This is the liberals' moral blind spot and amounts, I think, to their most grievous sin, because it is a betrayal of the very finest nation that has ever existed. It's worrisome, too, because while liberals seem to be losing political power, they retain guiding influence over cultural institutions ranging from universities to entertainment to news media. With that power, they are attempting to undermine American self-confidence, pride, and optimism—traits without which no civilization can survive.

MONA CHAREN is a syndicated columnist featured in more than two hundred newspapers. She also serves as a political analyst on television and radio public affairs programs and is the author of *Useful Idiots: How Liberals Got It Wrong in the Cold War and Still Blame America First.*

REED IRVINE

· ★ ·

FREE CHOICE VERSUS MAUDLIN
SENTIMENTALITY

As a teenager, under the tutelage of a radical high school history teacher, I was introduced to Marxism. I demonstrated that I had a heart by sympathizing with the poor and downtrodden and embracing socialism as the solution to the problem of poverty. This was not because my family was impoverished. My father was a laborer who kept his job throughout the Great Depression. My mother supplemented his income with earnings from substitute teaching. My three brothers and I all made a little money delivering newspapers from an early age, going on to better-paying jobs to put ourselves through college. Our sister, who was the oldest child, became a schoolteacher. Her income helped sustain our standard of living through the Depression years. We had few luxuries, but we never considered ourselves as poor.

The radical views that had been implanted by my high

school history teacher were cultivated by a few professors at the University of Utah. I carried them with me throughout my service in the navy and later the marine corps during the war. I never joined a communist organization, but I had a lot of sympathy and admiration for our Soviet allies. And that stuck with me when I studied economics in graduate school at the University of Washington and Oxford, where I was a Fulbright scholar.

After completing my studies at Oxford, I came to Washington and landed a job as an economist with the Board of Governors of the Federal Reserve System in the fall of 1951. This was at the height of "McCarthyism." Much is said about the reign of terror in Washington, with government employees losing their jobs because of their "liberal" views. My views were socialistic, which was by no means rare even among economists employed by the Federal Reserve.

But I was lucky. The director of the Division of International Finance, where I was employed, was a classical conservative economist, and his staff meetings were educational seminars unlike anything that I had encountered in college in the United States or at Oxford. My job involved keeping abreast of developments in the economies of countries in the Far East—primarily Japan, Korea (North and South), China, Taiwan, Hong Kong, the Philippines, and to a lesser degree Indochina, Malaysia, Singapore, Burma, Indonesia, Pakistan, and India.

I did a lot of writing, and I learned how important it was to be factually accurate. It soon became apparent to me that the more governments tried to control their economies with central planning, exchange controls, wage and price controls, unrealistic interest rates, tariffs, and so on, the worse the results for the people. I had access to reports from China and the Soviet Union that did not get widely distributed. What they revealed was that communism was a huge failure disguised by phony statistics. I wrote a paper on Mao Tse-Tung's "Great Leap Forward" in agricultural production, exposing it as just the opposite. The figures were phony, and the people were starving. It became known as "The Great Leap Backward." I also exposed that the Soviet Union's claims that their industrial production was growing faster than ours were based on phony figures, pointing out that their production was greatly exaggerated by counting goods that were defective and/or unwanted by consumers.

I saw that Hong Kong, the economy that came the closest to following Adam Smith's laissez-faire doctrine, ran circles around India, which was saddled with government-devised five-year plans. One of the things I learned when I finally got a chance to visit this astonishing British colony in 1957 was that economic statistics are not as important as economists think. I asked Douglas Cowperthwaite, the financial secretary, why they did not provide more than a few rudimentary statistics. His answer was that London was constantly asking for more, but he resisted their requests because if he provided them, he would be deluged with demands that he take this action or that.

I became a conservative by choice when I finally learned that free choice is a more important guide to action than maudlin sentimentality that encourages reliance on government intervention in managing economies. As a liberal, I had been guided mainly by emotion. That changed when I learned the importance of facts. At Accuracy in Media, we have found that journalists, most of whom are liberal, have an emotional attachment to facts and theories they have embraced. They are unwilling to consider, much less report, evidence that calls them into question.

REED IRVINE was the founder and chairman emeritus of Accuracy in Media (AIM), a nonprofit, grassroots watchdog of the news media. He served as editor of the *AIM Report,* wrote a weekly syndicated column, and was chairman of Accuracy in Academia.

ADAM H. PUTNAM

· ★ ·

THE GREATEST NATION WITH A
GREATER PROMISE

Churchill once said, "To be conservative at 20 is heartless and to be a liberal at 60 is plain idiocy." Far be it from me to criticize Sir Winston, but I grew up watching Walter Cronkite sign off by reminding Americans how many days the hostages in Iran had been held. In elementary school, we had moved beyond air-raid drills (because everyone, including students, knew how inadequate a school desk would be), but *Red Dawn, Mad Max,* and *The Morning After* made the nuclear winter a certainty and caused more than a few of us to peek out our classroom windows in search of Soviet paratroopers.

In short, these were scary times. If the Russians did not nuke us, the Japanese were going to buy us. Amidst that uncertainty, one man—Ronald Reagan—reminded us that we were the greatest nation with a greater promise. It was not about being a Republican or Democrat; all my people were

Democrats. You had to be to vote in a local election. But I heard in President Reagan what I saw in my hometown: America's potential is her people, not her government.

When families needed clothes, the Church Service Center was there. For food banks, you turned to the Key Club canned food drive. If there was a severely burned child, the Shriners cared for him. Need glasses and cannot afford them? Call the Lions Club. Each local charity and civic club identified a different need and attempted to fill it. While they did not solve every problem or cure every ill, these organizations helped many in the community, and with no law to compel them but the Higher One.

When Reagan called the Soviets an "evil empire," he was only saying what I'd heard many times before in Sunday school or church or at the kitchen table. When a nation abandons God, it is doomed to fail. Faith in the Almighty makes us stronger, not weaker as a people. The town that raised me allowed the mayor to lead a pregame prayer before high school football games, built monuments for fallen veterans of every war, and saved the fire engine that sounded the joyous alarm of V-J Day for parades. Faith and tradition were neither a source of shame nor a limit on progress; rather they were points of pride and the ties that bound one generation to the past and the future.

I saw the good government could do, and the limits of its ability. While in the legislature, I was offered two semi truckloads of potatoes from a farm share program. Being in government at the time, my first call was to the social services agency, whose personnel promptly told me they had no use for several tons of spuds. Even one of the largest charitable organizations could not handle that quantity of food. It was my secretary, the wife of a Baptist minister, who, in two days, rounded up church volunteers to unload and distribute the abundance of food to the people who needed it on short notice. It was a powerful lesson on the limits of even the most benevolent government.

Being conservative was more an affirmation for me than an evolution. No doubt, others growing up in the same town at the same time may find my own observations narrow or biased. But I have seen the lessons of my childhood writ large on the national stage and the conclusions are the same: our strength as a people is our people.

Helping different generations to connect and learn from one another enriches the young with the hard-learned lessons of the old, while the youth give their predecessors pride in having made the sacrifices they did and confirmation that it was for a brighter future.

CONGRESSMAN ADAM H. PUTNAM represents the Twelfth District of Florida. He serves on four committees: Agriculture, Budget, Government

Reform, and Joint Economic. He is the youngest member of Congress and also serves as chairman of the Subcommittee on Technology, Information Policy, Intergovernmental Relations and the Census, making him the youngest subcommittee chairman in the 108th Congress.

JACK OLIVER

· ★ ·

DEFINING SUCCESS FOR OURSELVES

I am a firm believer that God blesses each of us with many talents. America is the greatest nation on earth today because individuals are free to maximize those talents and create opportunities for success. And we have the liberty in America to define success for ourselves. For some, success is defined as being the best. For others, success is defined by what they help others do, like a teacher who helps students achieve their dreams.

Conservatism, to me, is the passionate defense of this American ideal. The ability to make the most of the rich blessings God has bestowed on us requires the complete freedom to make choices for ourselves and to shoulder the personal responsibility for these choices. It is only in this environment that success, however you define it, can be created from the simple spark of an idea, the pursuit of a dream, or the desire for a better life.

People, not government, are the best caretakers of freedom and opportunity. Certainly government has an essential role to play in this theater of our lives, but it should never get top billing. America has always been a place where people who are willing to work hard can accomplish great things for themselves, their families, and their communities.

Keeping this American dream alive for future generations is why I am a conservative.

JACK OLIVER was deputy finance chairman of the Bush-Cheney 2004 campaign and former RNC deputy chairman.

ZACH WAMP

· ★ ·

THE CORNERSTONES OF A HEALTHY SOCIETY

One of the fondest memories of my first term in Congress (the 104th Congress that elected the first Republican majority in forty years) was a presentation by our second session "class president" George Radanovich (R-CA) on the four cornerstones of a healthy society: the family, the church, the free enterprise system, and the government.

At the local, state, or federal level many agree that the role of government is to be limited and effective. The real challenge comes in resisting the pressure to allow the federal government to be the government of *first* resort instead of its constitutional place as the *last* resort. We should look to our families, our local communities, and our respective states before asking the federal government to meet any need not specified in the Constitution.

The first and perhaps the most important of these cornerstones is the family. Congressman Radanovich expressed the impact that the American family has on our culture and how the traditional family has served us well. The government should neither replace the family nor infringe on its role in our society. The family is under stress in America. It is under attack in many places around the world. If we are to be strong, we must strengthen the family unit. You may call it "old fashioned" or "conservative," but I call it common sense.

The church, under this definition, extends beyond denomination, sects, or even religion; it includes philanthropy and the groups that President Bush calls "armies of compassion." In his famous book, *Democracy in America,* Alexis de Tocqueville writes that the goodness in America is found in her people. This concept says a lot about the importance of faith-based institutions and the American heritage of giving more than we take. After all, Jesus said we are to love God with all our heart, soul, mind, and strength and treat everyone else the way that we like to be treated ourselves. Legendary golfer Chi Chi Rodriguez told me once that "takers eat well, but givers sleep well."

Finally, the American free enterprise system is truly "the goose that laid the golden egg." When we balanced the budget in the late 1990s, it was not a result of spending cuts or legislative belt tightening. The truth is the market economy of our great nation roared as a result of lower taxes, a supply-side approach, and entrepreneurial leadership in the information and communication sectors. Revenues finally outpaced expenses and magically the budget was balanced, creating misguided debates over what to do with "surpluses."

During the presentation nearly a decade ago, the "Class of 1994" raised eyebrows when we got to the fourth cornerstone of a healthy society—the government. Some conservatives believe in such limited government that they would not support funding for the new Department of Homeland Security or the interstate highway system through gas taxes. While this "libertarian approach" has constitutional merit, I respectfully submit that a true conservative should be neither a total libertarian nor a neoconservative. A good conservative will find a healthy balance in between. On the fiscal front, millions of Americans wanted the "government" to respond to the tragic events of September 11, 2001, and rightly so. The federal government must be efficient and effective while limited in size and scope; however, there comes a time when we need strong and decisive government action.

In regard to social issues and the government, my personal belief is that the Judeo-Christian teachings of the Bible are a fitting standard for public policy. The biblical worldview has stood the test of time. From its beginning, our nation has valued respect for the plurality of religious expression. Instead of attempting to stamp out all vestiges of religion from public life, the leaders of our nation should encourage the freedom to express our spiritual heritage.

Judeo-Christian values are a timeless and strong foundation for today's America.

When I became involved politically in the early 1980s, my primary motivation was that the federal government was trying to be all things to all people. I saw that the federal government had grown so much that it was trying to replace the traditional cornerstones of American society. The "Great Society" had created a more dependent people. The government needed to be reformed. In my opinion, the most significant change brought about by the Republican majority and the "Contract with America" was welfare reform. Today, millions and millions of Americans who were on welfare prior to 1996 enjoy the dignity of a job. The paradigm did shift and the reforms are still working all across America.

I am a conservative because I believe that the cornerstones of the family and the church can do more for needy people than a large federal government. But the "armies of compassion" must mobilize to do their mighty work without government interference or obstruction. We must streamline regulation, repeal unnecessary taxes, and eliminate frivolous litigation, so that the free enterprise system can provide the dynamic growth and opportunity for people willing to work, invest, and take risks.

These four cornerstone institutions provide the foundation for a just, compassionate, and prosperous American society.

CONGRESSMAN ZACH WAMP represents the Third District of Tennessee and serves on the House Appropriations Committee. He also serves as vice chairman of the Energy and Water Subcommittee and is on the Homeland Security Subcommittee and the Interior Appropriations Subcommittee.

PAUL RYAN

· ★ ·

GOVERNMENT DOES NOT KNOW BEST

I believe in the God-given potential of every individual and that freedom is the best environment for people to reach their potential. This is the underlying reason I am a conservative.

People should be free to exercise their minds and work to realize their dreams, as long as they don't trample on the rights of others in the process. Take the rights mentioned in the Declaration of Independence—the right to life, liberty, and the pursuit of happiness—and add the right to private property, and you have a good standard by which to judge the proper role of government. Government should serve those it represents by upholding these rights, not infringing on them.

In my view, government does not know best. Rather, free people tend to be more creative, more productive, and more efficient than the state in any number of areas. People have a great capacity for observing when certain needs in their community are unmet and can respond more quickly and effectively than government.

———————

CONGRESSMAN PAUL RYAN represents the First District of Wisconsin and is a member of the Committee on Ways and Means and the Joint Economic Committee.

THOMAS A. SCHATZ

· ★ ·

THE IMPORTANCE OF FISCAL CONSERVATISM

During my twenty-five professional years in Washington, I have watched government grow ever larger, even after Republicans took over both Congress and the White House in 2002. It is now apparent that there is only one governing body in the nation's capital, and it is a bipartisan spending party at the taxpayers' expense. That is one of many reasons why I am a conservative, fighting government waste, fraud, and abuse in order to prevent our children and grandchildren from inheriting an unsustainable burden of debt.

I did not always feel this strongly about fiscal mismanagement; in fact, like the greatest conservative of our time, Ronald Reagan, I was once a Democrat (but always concerned about wasteful spending). It soon became apparent that Democrats had no interest whatsoever in providing an efficient or limited government, and I joined a party that stands for lower taxes, strong defense, free trade, and judicial restraint.

Since President Reagan had established the President's Private Sector Survey on Cost Control, better known as the Grace Commission, which spawned Citizens Against Government Waste (CAGW), it was probably inevitable that my inherent fiscal conservatism would lead me to the Republican Party.

But that conservatism is also what allows CAGW, the organization I am honored to lead, to criticize both the Republicans and Democrats when they waste our tax dollars. In fact, the senator who has brought home the most pork-barrel spending per capita over the last three years is a Republican, Senate Appropriations Committee chairman Ted Stevens (R-Alaska). Of nearly equal notoriety is the long-time "King of Pork," Sen. Robert Byrd, Democrat of West Virginia. CAGW has been called Congress's "pork-barrel patrol." Our nonpartisan approach to cutting waste, along with our annual Congressional Pig Book, has helped make CAGW a leading national conservative organization.

As a conservative, I enjoy coming to work every day, trying to protect the hard-earned money of taxpayers today and prevent my children and grandchildren from bearing the burden of trillions of dollars in debt in the future. As our cofounder, J. Peter Grace said, "To advocate an efficient, sound, honest government is neither left wing nor right wing, it is just plain right."

THOMAS A. SCHATZ is president of Citizens Against Government Waste.

MICHAEL K. DEAVER

A REAGAN MAN

Many of my friends who relish the moniker "movement conservative" will no doubt be chagrined that I was asked to edit this collection of essays titled "Why I'm a Ronald Reagan Conservative." After all, along with James A. Baker III, Reagan's first White House chief of staff, I dodged a lot of bullets because I was often charged with the unpardonable sin of being a "pragmatist" in a conservative administration.

Until now, I never spent much time thinking about a label for my political beliefs. But one thing I know for sure, I always was—and always will be—a Reagan Republican.

I'm proud to say I've been a Reagan man since the early days, 1967 to be exact. Before that I put my heart and soul in the Barry Goldwater presidential campaign of 1964. At the time, the right was a caricature of itself, reaching for votes by reaching for the lowest common denominator. It was a party not of ideas, but of trepidation. John Birchers were alive and well, dominating its ranks at the grassroots, calling for the impeachment of the chief justice, booting the UN from our shores, and proclaiming that the dental use of fluoride was a commie plot to not only ruin our teeth, but to alter our minds.

Four out of five voters probably thought we were nuts.

Despite the nonsense bubbling up from some factions of the party, Goldwater overcame it with the right message. Sadly, he was the wrong guy to carry it forward. I signed up with Goldwater in 1964, drawn by his call for a return to the values and policies that made America the dominant power after World War II. In those early days of mass communication, he lacked the requisite tools to lead a majority party. Too strident, rigid, and sometimes just too mean, Barry Goldwater didn't sit well with the American public. He ended up carrying just six states against President Lyndon B. Johnson.

I woke up the day after the election vowing never to follow a candidate off a cliff again.

Two years later, as a field representative for the Republican State Central Committee of California, Ronald Reagan began campaigning for governor on a platform of what he called "Common Sense."

I had the great fortune of going to work for him in December of 1967 and later became one of his closest aides. It was during those years that I accepted Reagan's brand of "conservatism." It was just as muscular as the policies developed by Goldwater and company, but was tempered with a humaneness and compassion that had no home in the Goldwater campaign.

Reagan certainly had the firmness that I respected, but to this day, I have not met a more kind and gentle person. As America continues to learn more about the fortieth president, they will come to know the man I knew for nearly forty years, a man of keen intellect, an unyielding curiosity, and an unbending will.

History will record Reagan as the man who beat the communists without firing a missile, freed countless humans from the yoke of tyranny, and changed the way America looked at itself.

His views on free markets, taxation, the judiciary, and national security have inspired countless men and women to get in the arena of politics, building a new generation of Reaganites that will continue long after the people who served with him are gone. He did it all without ever personally attacking his opponents; he chose instead to simply out-maneuver them on the political battlefield.

Reagan was an ideologue, but he knew when it was time to negotiate. He had a pragmatic side, and as governor and president he'd roll up the sleeves and go toe to toe with the likes of Tip O'Neill and Dan Rostenkowski over tax reform. The secret to Reagan's success when dealing with the left was that he never took anything personally and he treated his adversaries with respect. Part of the reason he clicked with Gorbachev is that he respected the Soviet's point of view—too much so, some conservatives said at the time.

Contrast that approach with other conservative voices of my time. Growing up, I would listen to my father's angry rants about Roosevelt and the New Deal. My dad was a typical Republican man of his times, flustered by FDR's iron grip on political discourse and how he solidified Big Government in American politics seemingly in perpetuity.

It was not until Reagan came on to the national stage that we began to realize that those who believed in limited government and individual responsibility really had a voice in national politics. At the time, Reagan became a voice for a drifting, confused minority, and over the course of three decades he would convert his followers into a governing majority.

Perhaps this is the Gipper's greatest legacy. After all, how many men are there in human history whose name, when invoked, describes a revolution? Some may scoff at the word, but in modern America, thankfully, our revolutions are ones of ideas, persuasion, and values, not violence and upheaval.

The revolution that Reagan led is the reason why I am a Reagan conservative.

PETE SESSIONS

· ★ ·

BEAUTY AND FULFILLMENT

Every year, I hike with my family at the Boy Scouts of America's Philmont Scout Ranch in New Mexico. While not exactly mountain climbing in the most rugged sense, we nevertheless reach the top of Mount Baldy and, at an elevation of 12,441 feet above sea level, enjoy an unobstructed and spectacular view of New Mexico.

Away from the noise and distraction of news alerts, cell phones, faxes, and e-mail, my wife and I and our two boys take in this scenery with quiet appreciation. The view from the top of Mount Baldy at sundown has become fixed in my soul as an image that represents beauty. I have placed its picture in my mind to pop up in my consciousness whenever I think of what is truly beautiful in this world, a gift from our creator that expresses his power and majesty.

I hold that beauty dear today, but earlier visits to the same place did not produce in me the same appreciation of it. Discovering the beauty of the view from the top of the cliff was as much about me discovering what is important in life as it was about the mountain itself. It was up to me to find the beauty in the vista looking out from Mount Baldy during summertime sunsets, a personal journey that went beyond an annual hike to the top of a mountain.

I realize not everyone will have the same experience as I do when and if they visit Philmont. For many, beauty might not involve nature. Instead beauty might take the form of creation from man in art, music, dance, writing, architecture, or design. These creative expressions, and appreciations of them, are causes for celebration. Everyone's journey to discern, appreciate, and enjoy beauty is a personal passage every bit as significant as mine when I visit Philmont.

One's search for fulfillment and beauty is an individual quest. By definition, fulfillment and beauty can no more be dictated by an institution to someone for his or her appreciation and discernment than I can take a group to Mount Baldy and expect them to see the beauty that I see. This is a blessing: every man and woman's journey to discover what is beautiful is the life force behind our desire to be free.

How I have come to view beauty is unique; likewise, how I search for fulfillment and meaning in my life is equally unique to me. Both journeys are different from everyone else's, because I am different from everyone else.

Recognizing the personal nature of self-discovery in every person's search for fulfillment and beauty, I recognize the necessary sovereignty of the individual. Choosing to start at the individual for the basis of society, I am—by definition—a conservative.

Of course, as a conservative, I oppose excessive taxation, onerous regulation, and wasteful government spending. However, being a conservative does not begin with a reflexive "no." It begins with a celebratory "yes," yielding a path to every citizen's individual pursuit of fulfillment and beauty in the course of his or her life's journey. My opposition to burdensome taxation, regulation, and big government is a by-product of my desire to see everyone find fulfillment and beauty and my recognition that collectivist burdens historically get in the way of individuals freely pursuing their own journeys.

As an elected representative, I have come to believe that politics is the daily practice of lessons learned in history. Liberals, conservatives, and moderates debate every day how to interpret that history, and these disparate interpretations are the roots of most arguments in Congress.

Totalitarian societies have many common traits, but foremost among them is their near obsession of what forms beauty and fulfillment can "correctly" take in their respective cultures. Writers are thrown in the gulag for merely writing, the same for musicians playing certain types of music, dancers dancing, painters painting, and so on. Architecture and design are both forced into strictly defined molds. Dictatorial societies, whether they are communist, radical

Islamist, or Hitlerian fascists, produce hideous and distorted expressions of beauty and a warped path to fulfillment.

In response, freedom-loving peoples ultimately reject those ideologies—and so many have throughout history. Like all Americans, I enjoy the bounty and freedom that has made our nation the envy of the world. We should always remember that what makes our country unique is the right of every citizen to find his or her own expression of beauty and fulfillment.

Our strength as a nation is nothing more than the combined strength of individuals, as we come to terms with the creator and the challenges we face in our attempt to find meaning behind the monumental questions of life. There has never been a more dynamic society in history than the United States, because there has never before been a society that has placed so much trust and freedom in the people to discover for themselves the value and meaning in our fallible selves and mortality.

For these reasons, I am a conservative, and I believe that America's greatest days lie in the future.

May God bless you and your family in your discovery of beauty and your own fulfillment in this, the most free society ever conceived on Earth. Who knows, maybe we'll both share the sunset—or sunrise—from Mount Baldy one day. I hope so.

——————

CONGRESSMAN PETE SESSIONS represents the Thirty-Second District of Texas and serves on the House Select Committee on Homeland Security, where he sits on the Subcommittee on Emergency Preparedness and Response. He is vice chairman of the Subcommittee on Cybersecurity, Science, and Research and Development and also serves on the House Rules Committee.

DARRELL ISSA

· ★ ·

THE PILLARS OF OUR NATION

President Ronald Reagan, in his farewell address to the nation, fondly recounted the Reverend John Winthrop's image of the American land as a shining city upon a hill. Winthrop's 1630 vision of America was the first concept of America as a land of freedom and opportunity upon which the eyes of the world would focus.

As President Reagan told our nation, Winthrop's image of a land that would later become a nation has not lost relevance. The freedom, security, and economic opportunities of free enterprise that drew the Pilgrims to this land nearly four hundred years ago and later convinced the Founding Fathers of the need for American independence remain the pillars of our nation and the foundation of American conservatism.

Conservatism is a direction, not a destination. It is a constant effort to protect the ideals that make our nation the greatest on earth.

I call myself a conservative because I strive for the direction of the shining city upon a hill with the knowledge that this is not a goal one can reach, but one toward which all conservatives much strive. I believe in our free-market system, the freedoms guaranteed to us by our Constitution, and in a limited government that protects the safety of the people as well as the qualities of a truly free society.

CONGRESSMAN DARRELL ISSA represents the Forty-ninth District of California and serves on the House Energy and Commerce Committee. He is an electronics industry leader and immediate past chairman of the Consumer Electronics Association.

WALTER B. JONES

· ★ ·

RESTORING THE PRINCIPLES OF OUR PAST

When I came to Congress in 1994, I arrived in Washington as a recently converted Republican. I had spent the early years of my political career as a Democrat. As the years passed, I began to feel I had been left behind by the party of my father, a congressman before me, and therefore felt compelled to join the ranks of the GOP.

Since I came to Washington, those who learned of my decision to "cross the political aisle" have often asked for the reasoning behind my choice. The bottom line to why I am conservative is this: my party seeks to protect the future of America by restoring the principles of our past. Above all else, I believe in adhering to the Judeo-Christian principles upon which our nation was founded. Most conservatives will tell you they work hard to protect the biblical and constitutional rights of all Americans, civil liberties that are constantly being attacked by the liberal Left.

That being said, I value conservative ideals because they protect our finest asset: our children. Young people are the future of this great nation. Protecting the unborn is the first step in respecting the lives of the youngest generation.

Additionally, I strongly believe continued expansion of the federal government means continued expenses coming directly from the pockets of hardworking constituents. Our government is already far too large. I am proud to be a part of a political movement that works at reining in this kind of growth.

I believe that most everyone involved in politics has a good heart, but I also believe that conservatives work from a message of compassion. It is my privilege to be called to such work. I will continue to advocate these principles as long as the Lord calls me to do so.

CONGRESSMAN WALTER B. JONES represents North Carolina's Third District.

DR. MARVIN OLASKY

· ★ ·

GOD IS WISER THAN ME

I'm a conservative because I realized in my twenties, three decades ago, that I wasn't very smart.

Oh, I'll match up brightness credentials against anyone's. Sky-high SAT scores. A top chess player in high school. Graduated from Yale in three years. Ph.D. with super recommendations from impressed professors. Written lots of books. Yada yada.

But my brainy political ideas led me at age twenty-two, after several years of war and poverty protests, to join the Communist Party. My brainy ideas about marriage without fidelity led me to wed at age twenty-one and split at age twenty-three. My brainy ideas about religion led me to atheism.

This morning I read these words in Psalm 92: "How great are your works, O Lord! Your thoughts are very deep! The stupid man cannot know." At a certain point, God used my desire to know more to show me how little I knew.

For example, to satisfy a Ph.D. language requirement, I had to improve my Russian, and one evening, just for reading practice, I plucked from my bookcase a copy of the New Testament in Russian given to me two years before as a novelty item and never even opened. To my surprise, the words had the ring of truth. (It helped that I had to read very slowly.)

One more example: An assignment to teach a course in early American literature also helped, since my preparation involved reading . . . Puritan sermons. Those dead white males spoke truth. Later, the writings of C. S. Lewis and Francis Schaeffer showed me Christian brains at work, but brains that understood the importance of respecting a far greater intelligence than theirs or mine.

Why am I a conservative? Because I know that God is wiser than me, so it makes sense to follow the Bible rather than my own playbook. It also makes sense that a civilization based essentially (although imperfectly and sinfully) on biblical teaching would have wisdom that deserves respect. That's why I'm inclined to respect the wisdom of Western culture, earlier generations of Americans, and traditional institutions—family, church, business—instead of assuming that modernity knows best.

Theologian J. I. Packer has summarized the biblical message in three words, "God saves sinners"—and I can attest to how, changed by God's grace, I quickly left communism and slowly left utter self-centeredness. A marriage in 1976 that has only become stronger as the years have gone by, four sons who are all doing well, the ability to work productively and honorably as a writer, editor, professor, and elder— I know all of this comes from God, because I've seen the result of my own inclinations.

Many who have never had that born-again realization continue to be prideful. Those convinced of their own wisdom often are the worst: that's why professors often gravitate to the left. Those frequently kissed up to, like media stars and Supreme Court justices, may also head left. The liberal emphasis on liberation from traditional institutions assumes that we're bright enough to do much better. Liberals in power harass those institutions. Look at welfare's war on family, the Supreme Court's assault on unborn children, and the tens of thousands of regulations that hamstring the freedom to work.

I'm not blind to the heavy burdens that the word *conservative* carries. The Social Darwinist variety of conservatism—humanity evolves economically through survival of the financially fittest and elimination of the poor— turned its back on the needy in the past, and still sups with racism. And yet, the conservative movement of the past two decades, with its strong Christian saltiness, has relegated Social Darwinism to one small corner.

Conservatives generally understand that progress comes one by one from the inside out, not million by million from the top down. Most conservatives understand that all man-made institutions and all people are flawed, so utopia is not one revolution or one great leader around the corner. Conservatives, like liberals, like all of us, are sinners, but conservatives who privately defend sin at least do not try to use governmental force to push others to sin. Liberals, though, regularly purchase government-surplus stain removers that in practice grind the evil deeper into the social fabric.

That's why I'm a Christian conservative. "Claiming to be wise, they became fools," the book of Romans says of those who do not honor God. That was me. Now I claim to be dumb in relation to God—but not so dumb as to think myself wise.

DR. MARVIN OLASKY is editor in chief of *WORLD* magazine, a professor at the University of Texas at Austin, and a senior fellow of the Acton Institute.

SCOTT McINNIS

· ★ ·

COMMON SENSE

I choose to be a conservative because the ideology lends itself to a commonsense approach to governance. The most effective way to illustrate the strength of a given political philosophy is to compare and contrast opposing philosophies in their application to an individual issue.

Many Americans believe that all policy for the protection of the environment is conceived in liberalism. In fact, the liberal philosophy lends itself to a strict approach that disallows any use, be it recreational or the use of resources—a tendency that has led some to elevate trees and birds to a position where their protection negatively impacts humans. Conservatives opt for a more logical approach by considering the realistic needs of humans while not overlooking the importance of preserving the environment. They do this with the ultimate goal of finding a balance between the two.

This is the multiple-use approach to the utilization of our

public lands, a practice harkening back to the ideals of the father of American environmentalism—President Teddy Roosevelt.

The key to the successful implementation of this multidisciplinary practice is through the sound stewardship of these lands and our renewable resources. This means that we must ensure that the tools we use to harness our natural treasures do not cause irreparable harm to the environment or diminish our ability to reap the future benefits of a resource. Being an active steward of public land also means guarding our resources from impending harm, mandating that we act to reduce any threats to the health and stability of the environment.

Humanity can coexist in harmony with nature and utilize our planet's many resources for the betterment of our daily lives. The solution to ensuring the vitality of this relationship is for humanity to assume the responsibility to properly manage, protect, and utilize these natural treasures. History makes clear that a commonsense, conservative approach is the most effective way to achieve these goals. It is my belief that this principle is applicable to all political issues. That is why I choose to be a conservative.

———

SCOTT McINNIS is a U.S. representative representing Colorado's Third District and chair of the House Resources Subcommittee on Forests and Forest Health.

ROBERTA COMBS

· ★ ·

HISTORIC JUDEO-CHRISTIAN PRINCIPLES

Traditionally, there has been nothing that has identified a conservative more clearly than a belief system that encompasses the historic Judeo-Christian worldview as stated in the Bible. I believe and adhere to the principles of personal accountability and social responsibility that are evident therein. Conservatism did not command me to embrace a belief in the Holy Scriptures, but rather believing them has led me to become a conservative political activist.

My worldview has been shaped by my faith and belief in the teachings of Jesus Christ. It is clear that the Founding Fathers of this great country also held to his teachings. The foundational documents of the United States of America are reflective monuments to his compassion, his teachings of the immeasurable worth of all human beings, as well as his commands to value truth and justice above all things.

It is my personal belief that the stream of liberalism

within this country has eroded and denigrated the majesty of our constitutional government. I cannot embrace liberal thought or political action because I see it in direct conflict, not only to my faith, but to the security and sanctity of this nation.

To the end that I am held accountable by my faith and by my love of this country, I do my best to expose the destructive nature of the liberal agenda as it is practiced in the political arena. I feel it is my right and my duty as a grateful citizen of this country.

ROBERTA COMBS is president of the Christian Coalition of America, the largest conservative grassroots political organization.

JO ANN EMERSON

· ★ ·

STRONG VALUES

In my southern Missouri district, there is no sprawling metropolis, no big industry, and no international airport. What we do have, though, are hardworking, patriotic Americans.

Agriculture drives the small businesses of a hundred local economies. We grow everything from cotton to citrus. Families and churches are the foundation of our communities. I am a conservative because strong values reinforce those families and keep our communities growing.

Big government did not break the sod in Missouri, nor did it build the thriving agricultural economy, nor did it endow us with our deep faith. To me, conservatism means protecting individual rights from the infringement of big government, though capable federal government has its place.

Promoting and protecting free markets is an integral

duty of federal government. Trade has enabled the bounty of Missouri agricultural products to reach markets all over the world. Our charitable mission of food aid to impoverished countries is made possible by our global system of trade and humanitarian assistance.

Access to affordable prescription drugs for our seniors is another free-market issue in which I strongly believe. To apply downward pressure on the prices of prescription drugs, we must let free markets work.

And local control of our schools and municipal services is the best way to ensure that we get the highest possible return for our tax dollar. Burdensome regulations and bureaucratic inefficiencies only serve to dilute the purpose of our state and municipal institutions.

Through conservatism, we can accomplish those goals. In a future where government is respectful of the rights of the individual, a conservative philosophy will serve us well.

JO ANN EMERSON is the first Republican woman ever elected to Congress from Missouri and has served the state's Eighth District since 1996.

MICHAEL C. BURGESS, M.D.

· ★ ·

THE POWER OF THE INDIVIDUAL

I believe in the power of the individual. It is he or she who is ultimately responsible for his or her success or failure. In the United States, we are fortunate to live where we are given all of the tools necessary to become successful. We are provided opportunities unlike citizens of any other country, and with these opportunities come decisions and account-ability for each individual.

We should never look to the government as the first line of problem solvers. It is not the job of the government to run your life. That is your job. To that end, if government becomes so large as to handle all the country's problems, it will, in turn, become a source of problems.

Not only do I think that conservatives promote individu-alism through responsibility, I also believe that they pro-mote realistic optimism. In 1980, Ronald Reagan ran his entire campaign on realistic optimism. During his two terms in the White House, President Reagan brought forth global change through persistent, positive comments and a steadfast belief that individuals can make the right decisions when given the flexibility and responsibility to do so. He generated and promoted a climate where individuals would succeed, and now America still heralds independence into the twenty-first century.

———————

CONGRESSMAN MICHAEL C. BURGESS, M.D., represents Texas's Twenty-sixth District. He is also the only Texas Republican on the House Committee on Transportation and Infrastructure.

DAVID VITTER

INDIVIDUAL FREEDOM

I believe wholeheartedly in the sanctity and power of the individual, in individual freedom, and in the proven wisdom of peace through strength in foreign policy. I strive in my service as a public official to further these ideals and to embody the example of my favorite conservative leaders: Teddy Roosevelt and Ronald Reagan.

The time in my life when I knew I had become a true conservative was when I was studying in England as a Rhodes Scholar at Oxford. I saw firsthand during my time there how socialism had ruined an otherwise fine society, and it made me determined to work to protect the way of life that we enjoy in America.

DAVID VITTER of Louisiana is serving his first term as a U.S. senator.

TODD TIAHRT

· ★ ·

TAKE A RIGHT TURN:
My Conversion from a Democrat to a Conservative Republican

Yogi Berra said, "When you come to a fork in the road, take it."

Growing up on a farm in the Midwest, we saw politics like leftovers from a holiday meal. A majority of farmers were in their evening years and had vivid memories from the Great Depression. Many had experienced poverty, and some had even faced (and thankfully recovered from) bankruptcy. All of them remembered Franklin Delano Roosevelt with the fondness of their favorite radio show, for it was FDR who came to their rescue. He came for the little guy and gave him a second chance. They were Democrats, including my grandpa, John W. Steele. Modern politicians were still served up on FDR china like a fresh meal, regardless of stale and tasteless politics. They were political leftovers.

Just up the road was Meckling, a very small town,

trapped in a farming community. The hub of the community was Burton's Store. At Burton's was a converted house that was so small you felt like you were getting into a VW bug instead of a store. It had narrow aisles that were packed to the ceiling with dry goods and a few groceries. Pop was in the cooler up front next to the front door but way in the back was the bar.

After school, when I didn't have to catch the bus for a ride home, I went to Burton's. Everyone did. We stood outside and drank Pepsi, since it was only a nickel, and talked about cars and girls. The talk in the backroom bar was always loud and you could sometimes hear it from the street. Occasionally, it was about politics but never Republican politics. At Burton's they were never pleased about Republicans. They were Democrats, FDR Democrats, and that is just the way it was.

I was excited to turn eighteen. It was a real milestone for a young buck. Leaving high school at seventeen, nearly the youngest in my graduation class, left me wanting more freedom and responsibility. Announcing I was going to the courthouse to register to vote, my grandpa asked if he could ride along. Sure, I had driven Grandpa to the sale barn, where he bought cattle, since I was fourteen and knew he would make good company.

On the way to the courthouse we drove by the home place, where my mother was born. Grandpa reminded me that he lost that farm during the Great Depression. He lost quite a bit. He and his brother, Lee, owned more than four thousand acres, two elevators, a meatpacking plant, several buildings leased to a barber, a grocery, an ice house, and a small bank. It was the bank that toppled their financial tower.

When the banks failed in the 1930s, Lee and Johnnie, as they were known, tried to pay off all the people who had deposits in their bank. It took them under. Today, most people would have walked away from those who gave them their trust and declared bankruptcy. Lee and Johnnie gave a promise to repay, and as far as I know, they did just that. It took them awhile and they would accomplish that feat without the help of their old friends in the banking business.

I drove slowly down the gravel road past the home place.

Gazing at the home place, Grandpa said, "But it was a government man that came to the house after Jessie wrote FDR, and we got a loan for the farm that you grew up on."

Grandma must have been a persistent woman. I was told she wrote three times before the "government man" showed. FDR had come to the rescue and the political mold was cast. Grandpa would forever be an FDR Democrat, just like about everybody else in the area.

We parked next to the curb and slowly walked to the second floor of the courthouse. I registered as a Democrat and Grandpa smiled. I was feasting on leftover politics.

Being a Democrat wasn't too exciting. I went to college,

then went to college again, at a different address, got married to a wonderful woman, worked a couple of jobs and then as a result of my spouse's involvement in politics, decided to run for state representative in Kansas. Suddenly, after more than twenty years as a leftover Democrat, I asked myself, why?

My grandpa had made a compelling case on the way to the courthouse. Were the political parties served up on the same dinnerware today? Did the Democrats represent the little guy? Were they the party of second chances? Before I could file to run for office, I had to decide which party represented my heart, my hope, and my dreams.

The question that kept coming to me was like the front windshield of a car. A rearview mirror is an important part of driving. It lets you see what is behind you, like looking back into the past. But the biggest piece of glass is the windshield that lets you see where you're going, or where you will be in the future. Their importance seemed to be proportional to their size. Do I believe the country should look like it did when I became a Democrat, or can we build a better America? I was more concerned about where we were going and which political party would take us there.

Political parties usually do one of two things to pursue power. They either stand for principles or they bow to the political wind. My grandpa's party of FDR stood for the principle that Americans were great people; all they needed was a second chance, an opportunity to get back on their feet. Although economically America could not stand without the government during FDR's tenure, Grandpa did. And America certainly recovered from Pearl Harbor, moving the world toward freedom, sort of a second chance to make things right.

But I saw the modern-day Democrats differently. They seemed to have left the little guy's America, where anyone could make it, for a divided country. They pitted the African Americans against the whites. I found this unbelievable and unnecessary in a nation that struggles to be united and overcome the past.

Freedom seemed to be sacrificed in the party as well. Under the Democratic policies, political correctness limited free speech; gun control limited the right to bear arms, government regulations crept into our homes, cars, and work, limiting the control we have over our own lives. And higher taxes took away the ability to control some of our money. I was into freedom.

Then there was the dedication to the abortion industry. Planned Parenthood, National Organization of Women, and local abortionists aided and abetted the Democratic Party. I couldn't understand the cheapening of human life and still can't today. If we could take an unborn human's life, at any time during a pregnancy, even the ninth month, for any reason or whim, then what value would culture place on the living? Very little, and yet this was the official position of the Democratic Party. So, the Democratic Party actively pursued the dollars and support of the abortion industry and left behind most of America.

It seemed to me the "little guy," who was looking for a second chance in this country, was pushed into the backseat while the political winds blew the party's car off the road. The "little guy" would be taxed and regulated more by the Democratic philosophy and have less money in his pocket and less control of his life. The Left was in the driver's seat boasting of more government, more taxes, and less tolerance.

So I looked to the Republican Party. History tells us it is the party of Lincoln, who sacrificed greatly to abolish slavery and hold together a great nation. It was the party that fought for the thirteenth, fourteenth, and fifteenth Amendments to our Constitution that legally abolished slavery, gave African Americans the right to vote, and "life, liberty, and the right to own property." You have to wonder why more African Americans aren't Republicans.

CONGRESSMAN TODD TIAHRT represents the Fourth District of Kansas. He serves on the House Appropriations Committee and is also a deputy majority whip.

ROB SIMMONS

· ★ ·

THE FREEDOM TO DO WHAT IS RIGHT

I describe myself as "A fiscal conservative, because it's your money, and a social moderate, because it's your life."

Conservatives want people to be able to build their own lives and ensure the brightness of their own future, and only within a wide range of freedom is such a society possible. Government's responsibility is to create an atmosphere in which people, through their families, jobs, and civilizing institutions, can create and maintain such a society and to refrain from intruding more than is necessary to accomplish the task.

Free enterprise is the vehicle that enables people to flourish through their own industriousness. Where some see a patch of dirt, the entrepreneur envisions a garden. The products that make our lives easier, healthier, and more enjoyable were created by risk takers who had an idea and in-

WHY I AM A REAGAN CONSERVATIVE

vested to make it a reality. When we overtax these creative thinkers, we still the engines that grow our economy.

The burden of taxation takes away our freedom because it limits our opportunities to use our money as we wish. Too often after the monthly bills are paid, hardworking Americans do not have enough left to invest, save, and spend. Those are three essential components of our free economy. Whenever someone invests in a mutual fund, buys a new car, a new suit, or a new refrigerator, everyone in the economic chain benefits. When we deny families their disposable income through taxation, we limit their freedom to choose what they want to do and we diminish the economy.

It is necessary to make sure that government does not have more tax dollars than it needs to do the things it should do, such as defend our nation, provide assistance to those who are unable to care for themselves, and guarantee that every individual is able to exercise their constitutional rights.

Tax Freedom Day 2003 for American workers was April 19. That means it took people from January 1 to April 19 to earn enough money to pay their taxes. Although Congress has lowered a variety of taxes and eliminated a number of them, the tax burden remains too high and the tax code is still too complex. People need to be able to keep more of their hard-earned dollars because it's good for the economy and "because it's your money."

Freedom is not reserved for Westerners alone. At Tiananmen Square, in June 1989, thousands of Chinese risked their lives for the ideals upon which the United States was founded. The desire to make life better for oneself and one's family and to live in a free and orderly society where one does not tremble in fear is universal. Orderly freedom requires individual effort.

Samuel Adams warned, "Neither the wisest constitution nor the wisest laws will secure the liberty and happiness of a people whose manners are universally corrupt."

The responsibility of a civil society rests on the shoulders of the citizenry. Nevertheless, the government has a critical role to play in maintaining order.

In a free society government can create an atmosphere in which our civilizing institutions can do their work to encourage moral behavior. Lord Acton reminded us that ordered liberty is not the liberty to do what one wants but to do what is right. Self-government works only when people are prepared to govern themselves.

That is why teaching our children, at home and at school, right from wrong is imperative. We expect our schools to promote clear reflection, self-discipline, and public spiritedness, all of which contribute to a self-motivated, civil society. It is equally important that government pass laws that benefit families and respect the worth of the individual. Character development is essential because it alone ensures freedom.

People ruled by their passions and emotions are incapable

of self-government because they will be unable to conduct themselves in a manner conducive to ordered liberty. When accepting the social policy "it's your life," the individual must also accept that attached to the broad range of freedoms the government guarantees are a broad range of obligations that are permanently and inflexibly attached to those rights.

Conservatives understand that the more centralized a government becomes, the more apt it is to force the will of the few upon the many and thereby violate the rights it is supposed to secure. Government is just only when it exercises its fundamental duties, to guarantee individual rights and punish those who would deny people those rights.

Warning against the growth of the federal government at the expense of the states, Thomas Jefferson wrote, in 1800, that the states should be "independent as to everything within themselves, and united as to everything respecting foreign nations."

Although our nation is much different than it was in 1800, the theory remains applicable. For more than two hundred years America has walked the tightrope of rights, balancing how much freedom is possible before chaos results against how much government oversight is necessary before oppression results. We have walked that high wire better than any nation in history. Conservatives have no doubt that orderly freedom, free enterprise, and personal responsibility are inseparable; neither do they doubt the legitimacy of the historic principles that made us a great nation. As long as we are willing to protect these principles and adhere to them in our deliberations, America will remain the beacon of freedom and hope for the rest of the world.

CONGRESSMAN ROB SIMMONS represents the Second District of Connecticut.

ACKNOWLEDGMENTS

I'd like to thank my fellow Reaganites from home and abroad for their compelling submissions. I must also thank those individuals who wrote pieces that we were unable to include due to lack of space.

Since I started this book, two great conservative legends have passed. Ronald Reagan, the man who changed the world in his vision, and Robert Bartley, longtime editorial page editor of the *Wall Street Journal,* the one-man intellectual arsenal for the conservative movement. They will be missed.

Finally, I thank my co-editor Jeff Surrell for his time, effort, and insight.

BOOKS BY MICHAEL K. DEAVER

WHY I AM A REAGAN CONSERVATIVE

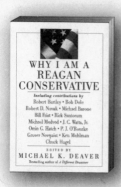

ISBN 0-06-055977-2 (paperback)

Former Ronald Reagan advisor and bestselling author Michael K. Deaver gathers the nation's leading figures to define conservatism, explore Reagan's impact on conservative philosophy, and examine what it means to be a conservative in America today. This timely and unprecedented new collection will elucidate and explore the impact of the man who fueled the fire of a rapidly growing political movement.

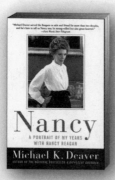

NANCY
A Portrait of My Years with Nancy Reagan

ISBN 0-06-078095-9 (paperback)

She was the daughter of a single mother, a Hollywood movie star, the wife of one of the greatest presidents of the twentieth century, a cancer survivor. And she waged her greatest battle against her husband's Alzheimer's disease. Nancy Davis Reagan has led an extraordinary life. Now, Mike Deaver, whose relationship with Mrs. Reagan dates back to the 1960s, shares the side of Nancy that only her intimates know.

A DIFFERENT DRUMMER
My Thirty Years with Ronald Reagan

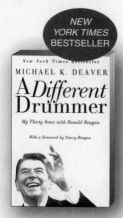

ISBN 0-06-095757-3 (paperback)

Deaver writes of the Reagan he has known: a man who was shy and deplored talking about himself, who would rather spend a party talking to a laborer than policy wonks; a man whose convictions remained unchanged over the course of his life, who never used pollsters to decide his position on issues; a man whose idea of relaxation was riding a horse, fixing fence posts, and chopping wood until his muscles ached and his hands blistered.